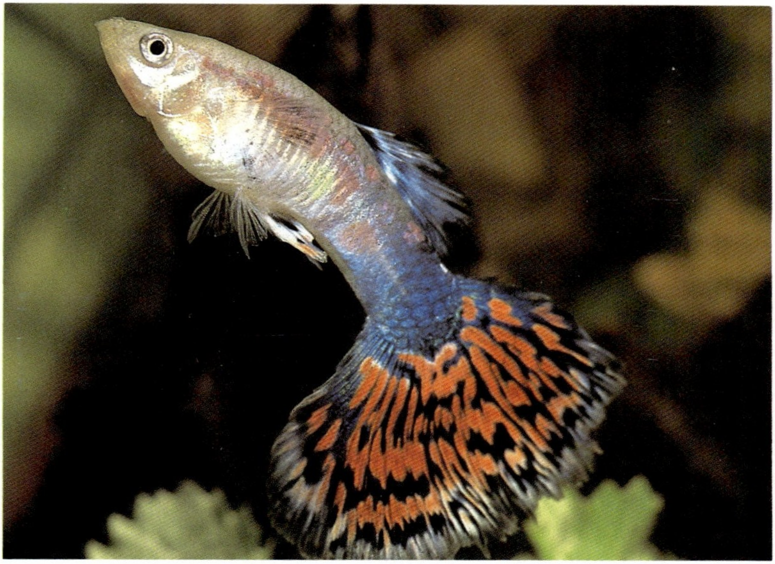

t.f.h.

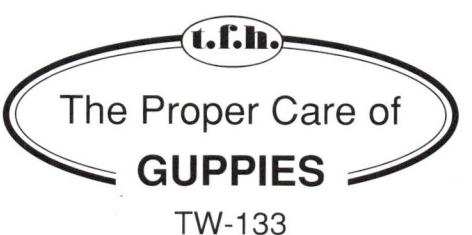

The Proper Care of

GUPPIES

TW-133

Title Page: *Fancy guppies with their vivid colors and elegant finnage are true aristocrats of the aquarium.* ***Photos:*** *J. Alderson, Aquarium Pharmaceuticals, Dr. H. R. Axelrod, M. Batell, Dr. M. Dulin, Dr. W. Foerster, I. Francais, R. K. Frickhinger, Gan Fish Farm, M. Gilroy, Dr. R. J. Goldstein, J. Greenleaf, Dr. H. Grier, N. Iwasaki, B. Kahl, K. Knaack, D. Kobayashi, H. Kyselov, G. Lewbart, C. Masters, K. Murakami, MC & P Piednoir, H. Richter, A. Roth, R. Schreiber, E. Shubel, S. Shubel, K. Tanaka, D. Untergasser, E. Wiatzka, R. Zukal.*

© Copyright 1995 by TFH Publications, Inc.

Distributed in the UNITED STATES to the Pet Trade by T.F.H. Publications, Inc., One T.F.H. Plaza, Neptune City, NJ 07753; distributed in the UNITED STATES to the Bookstore and Library Trade by National Book Network, Inc. 4720 Boston Way, Lanham MD 20706; in CANADA to the Pet Trade by H & L Pet Supplies Inc., 27 Kingston Crescent, Kitchener, Ontario N2B 2T6; Rolf C. Hagen Ltd., 3225 Sartelon Street, Montreal 382 Quebec; in CANADA to the Book Trade by Macmillan of Canada (A Division of Canada Publishing Corporation), 164 Commander Boulevard, Agincourt, Ontario M1S 3C7; in ENGLAND by T.F.H. Publications, PO Box 15, Waterlooville PO7 6BQ; in AUSTRALIA AND THE SOUTH PACIFIC by T.F.H. (Australia), Pty. Ltd., Box 149, Brookvale 2100 N.S.W., Australia; in NEW ZEALAND by Brooklands Aquarium Ltd., 5 McGiven Drive, New Plymouth, RD1 New Zealand; in the PHILIPPINES by Bio-Research, 5 Lippay Street, San Lorenzo Village, Makati, Rizal; in SOUTH AFRICA by Multipet Pty. Ltd., P.O. Box 35347, Northway, 4065, South Africa. Published by T.F.H. Publications, Inc. Manufactured in the United States of America by T.F.H. Publications, Inc.

t.f.h.

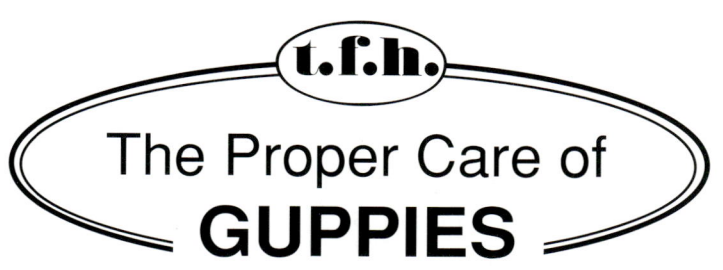

The Proper Care of GUPPIES

Stan Shubel

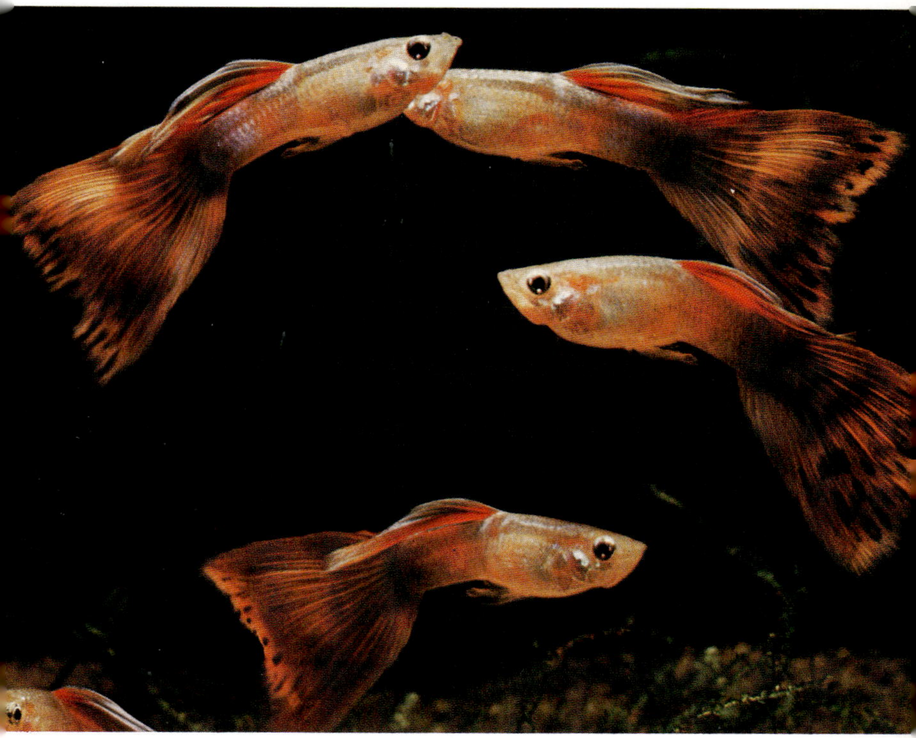

Contents

Foreword 7
Introduction 13
Equipment 21
Aquarium Maintenance 49
Feeding Your Guppies 77
Breeding Your Guppies 97
Showing Your Guppies 159
Guppy Health 187
Profit in Raising Guppies 239
Suggested Reading 252
Index 253

FOREWORD

Stan Shubel. **Facing Page:** *One of the better green deltas bred by Jamie Magnifico. Note how well the fish displays itself. The overall proportion and color are very good. This is a good fish for breeding or showing.*

Foreword

Writing a book does not come easy to me. Basically I have been, and still am a question-and-answer type person—and those questions and answers have been mostly verbal, not written. My introduction to guppies came around 1941. A small store near our home in Detroit, Michigan had a tank of fish in the window. Every time I walked by the store on the way to school, I would stop and watch the fish for a while. One day the owner was doing some work outside and I asked him what kind of fish they were. Of course, they were guppies.

The time I spent watching those fish must have left a permanent imprint on my mind. Some

years later, after serving a few years in the Army, I returned home to find my wife's aunt had a tank full of the same fish in her living room. She was only too happy to give me a jar full of guppies to take home. The next day I was out buying my first fish tank. The fish lasted about a week before I managed to wipe them out. Needless to say, within a day or two I was back to the pet shop for some more fish. I managed to keep these alive and once they started dropping young, it was necessary to have some more tanks. One fact I have noted about fishkeeping is that as the fish multiply, so do the number of tanks necessary to house them. At that time, there was very little information available on guppies, but for over 30

This is a variegated snakeskin bred by Frank Orteca. The fish has good color and good dorsal and caudal match, but the snake pattern is only fair.

Paul Hahnel and Stan Shubel at a guppy show back in the 60's.

years I have been working with guppies, and after countless hours of frustration and eventual enlightenment, some interesting findings have emerged. I'd like to share these with other hobbyists. These findings will be outlined in the various chapters and will deal with practical ways of breeding better fish as well as their care and upkeep. While reading through this book you will not find any reference material noted. I felt it was more important to report my own findings rather than to rely on other data that may or may not be pertinent, or even just a repeat of what has been written before. While it is useful to have a basic knowledge of genetics and water chemistry, the ability to observe the development of the fish and determine what characteristics go into the makeup of good specimens is more important.

I have been fortunate in knowing many of the early guppy breeders, such as Paul Hahnel, William Sternke, Larry Konig, Henry Kaufman, and others. Each of them has contributed something to the guppy hobby. In discussions, most of them were very sharp in one or two areas of fish raising. But when the conversation got out of the realm of their expertise, they did not have all the answers either. Today there are a few individuals around whose knowledge is equal to—and in some cases surpasses—these guppy greats of the past. I say this not to take away any of the credit from the pioneers of the fancy guppy but simply to point out that today's breeders also deserve their just due. And doubtless, these people will be the guppy greats of the future.

As you read through this

Bottom sword bred by Don Saurers. Dorsal and caudal shape fair with lack of color in front part of dorsal. Also lacks good color match between the body and the finnage.

The wild-type guppy, Poecilia reticulata.

book, keep an open mind. What I discuss is not carved in stone, but many of yesterday's and today's show winners are using these very same methods for success. By the same token, some of the information presented here is basic, common knowledge, but the basics are always, and shall always be the beginning of any successful venture.

To you, the hobbyist, and to your great fish.

–Stan Shubel

A trio of wild guppies.

Introduction

The guppy, or *Poecilia reticulata* as he is known scientifically, has been available to the aquarium hobby for over one hundred years. The guppy is collected from various locations in South America and the islands of Barbados and Trinidad. The wild guppy brought to America and Europe, however, bears little resemblance to the fish we see in pet shops and fish shows today.

After their initial introduction, guppies quickly became very popular because of their

Facing Page: *A tankful of male blue guppies. The best of these will be chosen for breeding.* **Above:** *Iridescent blue delta, a nice fish, but without breeding potential because of the extended ray on top of the caudal.*

brilliant colors and the fact that they were livebearers. Not much change in the wild stock was noted for a number of years until a Dr.

Ab on the East Coast provided some common wild guppies with good tank conditions and feeding methods. His fish grew much larger in size than usual and people assumed that he had discovered a giant strain of guppy. However, when these same fish were cared for in the normal manner of that day, they quickly reverted back to the small, wild-type fish. Shortly after Dr. Ab's observations, a Mr. Paul Hahnel came into the picture. He not only provided his fish with the necessary tank and feeding requirements for good size, he also set up a breeding program. He may not have known all that much about genetics, but he did have a great eye for good fish.

By allowing the better males to breed the females, in a few generations he was able to put larger tails on the males. Thus, by controlled rather than natural selection, he was able to produce the first truly fancy-tailed guppy.

Paul Hahnel became known as the father of the fancy guppy. His fish were what we term today veiltail guppies. As his stock was distributed, other breeders developed different colors and tail shapes from his fish.

In Florida, at about the same time, William Sternke and his wife were working on strains of their own. Their fish seemed to have slightly better color, but the caudal fins were nowhere near as even as the caudals on Paul's fish. Some very large-tailed males were also being produced on the East Coast by Frank Alger.

The Europeans had also been raising guppies for a number of years, but until some of the American wide-

INTRODUCTION 15

Above: A young black male bred by Don Saurers showing good body and caudal color. Dorsal and caudal are not in good shape. The dorsal is also off color. **Below:** A top sword bred by Don Saurers. The dorsal and caudal have a fair shape with lack of color on the front part of the dorsal. The color match between the body and the fins is not very good.

tail fish arrived there, they had only the smaller tailed varieties. Even today, in their judging, the Europeans use slightly different show standards from those accepted by the International Fancy Guppy Association.

In the U.S., the American Guppy Association (AGA) was formed with Larry Konig acting as chairman. This group did a lot toward getting some show standards set up. Larry also distributed a number of

Red female line with good caudal shape and color. She will drop her first young in about a week. **Facing page:** A show tank full of mixed guppy types.

As a rule, the European fish are smaller in overall size than the American type. Most of the size disparity is due to the difference in feeding practices and show standards required for delta-tailed fish to various guppy clubs throughout the country.

After the AGA faded out of sight, a group of guppy people met in Ohio to start up another organization called COGS or Congress of

Above: Half black red female. One of my breeders. Note the nice color blend in the caudal with the matching dorsal. **Below:** Young green delta. Once in a while, a few of these will pop out of one of my blue lines. Peak on top of the dorsal is not desirable.

Guppy Groups. The following year, the name was changed to the present International Fancy Guppy Association (IFGA). Since that time, we have seen a great change in the development of the guppy and vast improvements in judging standards.

leave you with a pretty fair idea of what the judges are looking for in show-quality fish.

This book is targeted for those people who are interested in breeding and raising show-quality guppies. It is not necessary to have a dozen or a

A group of blue deltas about six months old from my blue line.

Most of the photos in this book can be used to give you a good idea of what types of fish are being raised at this time. The brief descriptions given will point out both the good and bad points of the fish. The photos and their descriptions will

hundred tanks to enjoy this popular fish. If you have a spare tank or two and are bored with raising the easier types of fish such as discus or African cichlids, switch over to guppies and I guarantee you won't ever be bored with your fish again!

Above: Half black blue male. *Below:* Half black blue female.

Equipment

Guppies are often kept or bred in sterile, unfurnished aquaria. This is the technique used by professional guppy breeders whose focus is not "Aquarium Beautiful," but the fish and their bloodlines. The professional breeder constantly tends and supervises the condition of his fish. This supervision and regular water changes have no doubt led to their being kept in this way by professional guppy breeders. However, *any* professional fish breeder, of angelfish or oscars or guppies, is going to keep his fish in the barest possible aquarium because, let's face it, professional fish

Group of 2 1/2-month-old half black reds. Note the good dorsal shape and color match even at this age.

breeding is a different ball game. To the fish breeder beauty is gravid females, sterile, bare aquaria, and brimming vats full of brine shrimp nauplii—and beautiful fish.

THE AQUARIUM

Whether you are shooting for a decorator aquarium or banks of aquaria with generations of high-quality broodstock, you are going to need a vessel that reliably contains the water so necessary for their lives.

Show Tanks

An attractively planted aquarium with a school of perky, healthy guppies and even other species of compatible fishes is not difficult to set up. On the contrary. If you are primarily interested in keeping guppies rather than breeding them, it is best to

Dark blue delta. This fish was over a year old when this picture was taken. Note the length of the matching dorsal. This is still a very active fish.

Above: *Half black blue female. A good female is very nearly as colorful as her male counterpart.* ***Below:*** *A quality aquarium will give you many years of reliable service. Photo courtesy of California Aquarium Supply Co.*

keep them in a fully furnished, beautifully decorated aquarium. You will soon see that a decorated guppy show tank is little more work and certainly much more pleasing to the eye than a sterile keeping tank. It is this pleasure that helps ensure that the fish get the attention they need from the busy hobbyist.

Breeding Tanks

If, however, you are *serious* about breeding high-quality show guppies, there are certain things you must do. Depending on how many lines, or strains, you intend raising, you need a minimum of six tanks for each line. The 5- to 10-gallon tanks are best for breeding and raising young fish. For older and larger fish, the 15-gallon tank seems to work best for me. Over the years I have tried various larger sizes but have always come back to

The Aquarium Plant Plug contains an ideal growth medium for aquatic plants. Photo courtesy of Aquarium Products.

EQUIPMENT

There are many products on the market that will help you adjust your pH to the proper levels for your guppies. Photo courtesy of Aquarium Pharmaceuticals.

the 15-gallon size. You may find that you can raise your guppies just as well in larger tanks. That's fine; use whichever size tank that works best for you. One of the advantages of the smaller tank is the ease of handling when it comes to moving or cleaning them.

Once you've wrestled a 50-gallon tank around and tried to place it back on a rack after cleaning it, you'll come to appreciate the smaller sizes.

Glass Aquaria

Probably the all-glass aquarium is the most practical type to use,

You can purchase water conditioners at your pet shop. Photo courtesy of Aquarium Pharmaceuticals.

superior to the old-style stainless steel-framed tanks many of us remember. They are light and easy to clean in the modest sizes, and as a rule they are moderately priced. The only disadvantage I can see is that you do have to be a little careful in handling them to prevent damage to the edges. Also, care must be taken to see that they are resting on a level surface. Another advantage is that you are able to drain them and keep them dry for a long period of time without the fear of having them leak when you refill them at a later date—and that's something you could never do with stainless-framed tanks.

Acrylic Aquaria

Acrylic aquaria are gaining market share, and well they should. They are very light and easy to move. Acrylic tanks are available in some very interesting shapes for those of us who want a real showcase for some prize guppies in a beautiful decorator aquarium. On the down side, however, one must be a little careful not to scratch the acrylic during cleaning and handling.

BOWLS

One piece of equipment I find indispensable is the 1-gallon drum bowl. They have a number of uses for the guppy hobbyist and breeder. They will hold baby fish until you have a tank ready for them. They are handy for getting a good look at potential breeders, for separating females ready to drop their young, or even isolating sick fish. From a practical standpoint, it is much easier to medicate a small bowl of water for the treatment of a diseased fish than a whole tank—especially since many of the medications will destroy the bacterial balance of the tank.

Power filters are an excellent choice for show tanks. Photo courtesy of E.G. Danner.

Once in a while, when I run out of available tank space for baby fish, I will set up a gallon drum bowl with a small filter in it. You can keep young fish in this type of arrangement for a week or two with no ill effects. I did run into problems when

I put pregnant females in the same setup. It seemed as though as long as there were no places for the young to hide, the females didn't bother them. However, when the young darted behind the small filter in the bowl, the females would go after them when they came out. I tried this several times with the same results even though there were plenty of live baby brine shrimp in the tank. It appears that situations such as this trigger off natural responses that are normally somewhat controlled. I have also added plants when using the drum bowls to hold gravid females, but with only fair results.

RACKS

When planning your fishroom, try to figure out how many tanks you will have. Are they going to be set endwise or lengthwise, and how many rows of them will there be? Leave enough room for working in the tanks between each level. I have three rows of tanks, the top row being at eye level and the bottom row at about knee level. I don't really like to have the bottom row that low, but it was necessary to arrange them in this order to get enough tanks in a small area. Just below my mid-level row I have run a $1\ 1/2$-

Water conditioners can be used to treat water in a newly set up tank to make it safer for the fish it is intended to house. Photo courtesy of Aquarium Products.

EQUIPMENT

This is the left side of my fishroom. Note the racks, simply made of 2 x 4 lumber. The timer that controls the lights is on the far wall. Electric baseboard heaters are used to maintain a constant temperature level. The small black fan on the right side pulls in fresh air from an outside room. As you will note, there are quite a few fish in the tanks.

inch PVC pipe the length of the row with an inverted opening every few feet for draining my tanks without having to carry water. In a normal fishroom this drain pipe would be below the bottom tier of tanks, draining into a floor drain; but as I have to pump water *up* from my basement into a septic tank, I have to do it a little differently.

The racks in my fishroom are made out of lumber 2 x 4s. Use a good grade of pressure-treated lumber to resist water damage. If you do build your own racks, make sure you have enough supports so the rails don't sag. Some guppy breeders use angle-iron racks and others have developed their own custom designs. It's a matter of personal preference and pocketbook.

EQUIPMENT

Many of the guppies you will find commercially, like these, are produced in Singapore and imported in bulk from there.

LIGHTING

Lighting is an important factor in the fishroom. If you plan to raise plants in your tanks, you will need a light over each tank or a string of lights over each row. Basically, there are two different types of light to use, fluorescent and incandescent. As an individual type of tank light, the incandescent provides a good plant growing light. It has a couple of disadvantages though, one being that it generates quite a bit of heat and is quite costly to run for the amount of light that it produces. Although the bulbs are less expensive to buy, they do not last as long as the fluorescent type.

The fluorescent fixtures are more expensive initially, but last longer, and are cooler and cheaper to run. My own lights run down the center of the fishroom and are controlled by a timer. The lights are on for approximately twelve hours a day. They are on in the morning for about three hours during feeding, shut off during the day, and come on again just before I come home in the evening. They remain on for about an hour after the last feeding.

A small light is left on under the shrimp hatchers. It warms the water slightly, usually keeping the water 80°F, but its main function is to provide a faint light source which partially illuminates the fishroom, so that it is not completely dark when the overhead lights go off. The fish in this semi-light do not lay on the bottom of the tank, but rather stay suspended in the mid area, which is healthier for the big-tailed males. The "nite-light" also prevents too much of a shock to the fish when the overhead lights come back on. The fright sometimes causes them to jump out of the tank or run into the sides.

The overhead fluorescent lamps themselves, alternately placed, utilize warm white and cool white bulbs. You may wish to use a warm white to bring out the reds, or possibly a daylight bulb for blues or greens. I would advise saying away from the "grow" bulbs for your fish, especially under extended usage conditions. The number of bulbs needed will be determined by how bright you want your fishroom to be and whether or not you plan on raising plants. Fish can be raised successfully in very dim or very bright conditions. I have tried both and found that I had the best results using moderate lighting.

BACKGROUND

The walls of my fishroom are paneled in a medium brown wood grain color, mostly because inexpensive panelling is less costly than drywall. You can experiment with different backgrounds to see which

Carbon is a valuable filter aid. You can remove medications and toxins like ammonia from the water by putting some carbon in your filter overnight. It becomes saturated fairly quickly, but is ideal for a quick clean-up. Photo courtesy of Hagen.

Purple delta.

brings out the best color in your fish. The bottoms of all my tanks are painted on the outside with flat black spray paint. This helps to deepen the color of the fish and to make them feel secure. The bottom of an unpainted tank sometimes acts like a mirror, giving the fish no reference point and thus keeping them in a state of fright.

FILTRATION

At one time or another, I have tried almost all of the different methods of tank filtration for raising guppies. While it is possible to keep guppies with no external air supply or any type of filtration other than occasionally draining a portion of the water and replacing it with fresh water, it is not a practical

EQUIPMENT

path to follow. We will assume that you will have an air supply and some form of filtration. Practically speaking, there are four types of filters: outside box, inside box, undergravel, and sponge. Oh sure, there are some fancy types available, but for multiple tank setups they are impractical in both the physical and economical aspects.

Outside Box Filter

The outside box filter consists of a container for holding your carbon and/or marbles and filter floss. A siphon tube goes into your tank. An airlift or discharge tube runs the filtered water back into the tank after going through the siphon tube and through the filter. If using an air pump to drive this type of filter, you are also adding oxygen at the same time. If you are using the direct drive or motor driven outside filter, you are usually filtering the water

Young black female bred by Bill Orth. Very good caudal and body color, good intense black, but the dorsal is a little light.

In the Far East, this guppy would probably be called a blue cobra. It is somewhat similar to the snakeskin type.

faster, but are adding much less oxygen, other than through the surface flow action. This outside box filter is quite acceptable. The only drawback is that if the water level in the tank drops, the filter will lose its siphon and the filter will stop. The motor may be damaged as well.

Undergravel Filter

The undergravel filter works primarily through bacterial action with an external air supply. The undergravel filter is useful in keeping the water crystal clear in display tanks which is fine for most other fish. But for guppies, in my opinion, it is almost useless. After a week or so in an undergravel filtered tank, you can almost watch the males' tails fall apart. For a fish store they may be useful, but for the serious guppy breeder, they are not.

A good thermometer is essential when keeping tropical fish. Photo courtesy of Hagen.

Sponge Filter

Sponge filters are basically just a sponge with a combined inside siphon and air tube. Sponge filters utilize the same basic bacterial action as the undergravel filters and as a bonus are home to rotifers that the young fish like to nibble on. The disadvantage of the sponge filter is that it does not pick up the mulm from the bottom of the tank, and will occasionally cause the bacteria level to get out of control, possibly causing tail damage to the males. Some of the breeders use them with fair results. Mine were always clogging up. Try them. They may work for you.

Submersible heaters are wonderful. You can set them and forget them. Place the heater close to the bottom of the tank so you don't have to worry about it when you change the water. Photo courtesy of Hagen.

EQUIPMENT

A tank of young blue deltas. This is the type of inside corner filter that I use in all my tanks.

Inside Box Filter

Most breeders, myself included, use the inside box filter. It is simply a small box that sits inside your tank. The filter medium is usually filter floss and carbon or marbles. An airline is run down to the filter which, in turn, draws water down through the floss and bubbles the air back into the tank—again with a combined bacterial and filtering action.

Central Filtration

Another type of filter arrangement is the continuous loop from tank to tank to tank and then back to a central large filter. The only problem other than trying to maintain the water levels is that if there is a disease in one tank, soon there will be an epidemic in all the tanks. With the advent of the ultraviolet light for disease prevention, some improvement in this problem was noted, but I would still not recommend this system to the guppy breeder.

Filter Media

For many years, I used carbon as a filter medium, but it got to the point where I found it too expensive to use. I then went to marbles used in conjunction with the floss. The marbles were to hold the filter down. They don't keep the water quite as clear as the charcoal, but they don't cost anything after the initial purchase. The marbles don't give as much leeway in case of accidental overfeeding as carbon filtration. However, without going into a long discussion on the pros and cons of carbon usage, recharging and useful life, I believe it is safe to say that within a week's time under normal tank conditions, 90% of the carbon's effectiveness is gone. Carbon is a very porous material and collects all the debris and mulm that passes through the filter floss, thus causing it to clog up. You do continue to get some filtration as the surface becomes progressively coated, until the point is reached where about all you have left is a bacterial action filter.

Frequent water changes, regular replenishment of filtering material, and controlled feedings practically eliminate the need for carbon. One thing, while I am on the subject—*never* drain the tank and replace the filter on the same day. This, along with overfeeding, has probably messed up more good fish than every other factor combined.

The size of the filter should be in correct

EQUIPMENT 39

Above: Dark green delta bred by Jim Alderson. Iridescent green color in the caudal and body. The dorsal is a little off. He is a good fish other than lower edge of the caudal.

Below: Blue/green bicolor delta bred by Jim Alderson. Nice blend of colors in the caudal and dorsal, also a fair amount of body color.

proportion to the size of the tank. If the filter is too small, you will not have an adequate media area to both trap the particulate matter and to provide a sufficient bed for the required bacterial activity. A filter that is too large will take up valuable tank space and one that's too small won't do the job.

Drip Systems

Some guppy breeders are experimenting with drip systems. With this method, a continuous supply of fresh water at a controlled temperature is fed into the tank. A standpipe usually provides a runoff from each tank so it won't overflow. If you have the necessary plumbing and housing situation, this is perhaps one of the best ways to raise fish. You would still have to clean the insides of the tanks and the filters, but other than that, it would simplify your work a great

AOC (Any Other Color) bicolor delta bred by Jim Alderson. Good bi-coronation in caudal; dorsal is slightly off color and missing the caudal pattern.

EQUIPMENT

Purple delta bred by Jim Alderson. Nice large dark purple caudal with a fair matching dorsal.

deal. It would also enhance the growth potential of your fish and should eliminate the possibility of most disease problems.

Diatom Filters

When their tanks become cloudy, some hobbyists use a diatomaceous earth filter to clean up the tank. This type of filter will usually pump a hundred or more gallons of water in an hour and can quickly remove most of the suspended particles in the tank. Those who use them say they have few, if any, problems with disease transference.

HEATING

The use of individual heaters for a few tanks is acceptable, but if you are going to run many tanks you are better off with an enclosed room having a single heat source. You can use a small space heater (with proper ventilation and fire safeguards, of

This matched pair of red deltas is from a cross between two of my red lines. Being distantly related, there are none of the problems you might encounter with a hybrid cross.

course) or duct hot air into the room from your furnace. The only problem with this is maintaining a constant temperature. My own fishroom is heated by a thermostatically controlled 220–volt electric baseboard heater. Be sure that all the walls and ceilings are properly insulated. Once the tanks are up to the correct temperature it does not require too much heat to maintain that level.

With an enclosed room it may be necessary to bring in fresh air, as the humidity goes quite high. The best way I have found to accomplish this is through the use of a PVC line hooked to the input line of my air compressor. In the warmer months I draw fresh air in from the outside. When it turns cold I close off the outside inlet and draw air from the family room next to my fish room. Also, in the summer I use a small box fan about four inches square to bring more air in from the outside. This unit really makes my fish room livable in the hot summer months and keeps it smelling fairly fresh. Before I installed this fan in a small wall opening, heaters use 110 volts, while baseboard heaters use 220 volts. This makes them twice as efficient at putting out the same amount of heat. I would recommend using a thermostat mounted on an inside wall of your room. Place the actual heaters on the outside walls, preferably in an area

Gold albino delta bred by Frank Chang. This is a nice color variation from the usual color we see at fish shows. Good gold body with a fair color match in the dorsal and caudal. Both top and bottom edges of the caudal should be more even. The required pink or red eye color is easy to spot on the gold body.

the fish room at times was so muggy it was too miserable to work in.

From a purely monetary aspect, individual tank that is not directly under any tanks.

The temperature sometimes dips to 25°F below zero in my area.

Thankfully, the longest we have gone without electricity is three days with the temperature about 20°F. All I could do was keep the fishroom door closed during this time to keep the heat in. When the electricity finally came back on, the fishroom was down to about 60°F. All I lost during this potentially catastrophic occurrence was about a dozen fish. That is the value of insulation.

If you do use individual heaters, use caution when cleaning your tanks. It is easy to break the glass on the heater. If it is on when you do, you can get a nasty shock. I speak from experience. The same might be said about reflector lights on your tanks. One time my cat was playing around a tank and knocked the light into the water when she dipped her paw into the tank. That was about the highest I had ever seen her jump! She stayed out of the

Red albino delta bred by Frank Chang. Good overall matching color with a well-shaped body. The caudal is slightly short in relation to the body, whereas the dorsal is a little longer than the required three-to-one ratio.

Green delta bred by Jamie Magnifico. This is probably the best green guppy photo I have been able to take. It is a nice young fish with excellent color match in the finnage. Body color is also very good.

fishroom for quite a while. Thankfully, it wasn't a saltwater tank or she might have died.

AIR SUPPLY

Your air pump and compressor have to provide you with a constant oil-free supply of air in the most economical way. For a few tanks a pump such as a powerful vibrator may be sufficient. It is wise to purchase one large enough for expansion, because when you're raising guppies it seems that you are always adding more tanks. You will also need extra air to run your shrimp hatchers.

I have found it to be a good practice to keep my compressor on the floor of my fishroom with enough clear space around it so there is adequate circulation to provide for the cooling of the unit. I have seen where some fish people, in an attempt to cut down on the noise level, would place insulation around the pump and then wonder why the motor burned out.

Recently I purchased a regenerative blower which is ideal for aquarium use. The initial cost of the unit is fairly high, but when the high volume of clean air provided at low operational cost is taken into consideration, they prove to be quite attractive. The noise level is quite low, and the unit sounds something like a turbine. Because there is a cooling fan, a minimum of heat is produced. Back pressure in the lines seems to be the only critical factor to be taken into consideration. Also, with the blower unit, there will be no carbon build-up in the lines or valves—a definite plus. Use at least 1½-inch PVC pipe for the air supply line, preferably in a loop-type arrangement with holes drilled at required locations for the ⅛-inch insert air valves.

As you will be working with low pressure, you may need to run loop lines through a series of 3-way valves to your filters so that there will be little variance in output. You will find that it is possible to leave several valves running open and still maintain enough air for the filters. The open valves along with the large size should eliminate any back-pressure problems.

The PVC pipe is easy to work with. All you need is a hacksaw for cutting and the bonding glue. It is possible to do an entire fishroom air supply line in an hour or so.

The main air supply line should always run above your tanks. This prevents back-siphoning is there is ever a power failure. (This is also true when using any kind of air pump.)

EQUIPMENT 47

Above: *Multi veil bred by Frank Mormino. Caudal color distribution is acceptable. The dorsal match is close. The only weak point would be in the dorsal ratio. It should be at least another unit longer in length.*

Below: *Variegated snakeskin veil checking a small snail. Bred by Jamie Magnifico. Caudal pattern is good with slight weakness in the dorsal. The zebra bar pattern in the body is not a desirable characteristic and points would be deducted when judging the body.*

48 AQUARIUM MAINTENANCE

Aquarium Maintenance

SETTING UP

With most fishes, except for guppies, bettas, and killifish, it is necessary to have large tanks for proper growth. When raising guppies it is possible to produce good fish using only 5-gallon tanks. This makes it nice for those of us who have space problems. About the only drawback is the reduced number of fish you will have to work with. With a well-established line, it is possible to raise fewer fish and still compete well at shows.

Facing page top: Blue delta line with very good color and well-shaped body. There is a mismatch in the dorsal color, so this fish would not be used as a breeder.
Facing page bottom: Young red female with color highlights on fins. Once she has dropped young, she will almost double in size.

Add aged aquarium water and a used, fairly clean filter whenever you set up a new tank. Let the filter run for at least 24 hours after adding your plants. I use water sprite (*Ceratopteris thalicroides*) primarily and find that as long as the water sprite is growing well the tank is healthy for guppies. When the water sprite starts to die, tank conditions are deteriorating. Of course other plants can be used—in a display tank your only limits are availability and budget. One of my tricks when I do want to use plants is to use small glass containers and root the plant right in the glass. You can add a couple of these to a tank with no trouble and no gravel. Water sprite and swordplants do very well in this type of arrangement. No matter how practical I try to be, I do, from time to time succumb to the desire to see my fish swimming in a beautifully planted tank. In our drive to produce bigger and better fish, we sometimes forget to take the time just to sit back and enjoy our fish.

For the guppy breeder, gravel is anathema. Gravel is not your friend if it is your intention to raise or maintain show fish in any number. If your wish is to have a display tank with plants, gravel, rocks, etc., then I suggest a fine, round gravel. If the gravel is too coarse a lot of food will disappear into the substrate, where it will soon start to decompose, creating tremendous water-quality problems. Sharp-edged or pointed gravel is completely unsuitable as a substrate. The fish can easily injure themselves on

Young blue female with excellent body shape for a breeder. Good distribution of color in the caudal.

AQUARIUM MAINTENANCE 51

Half black delta. This is a nice, smooth young fish with good caudal/dorsal match and very good body shape.

it. The substrate must be rinsed very thoroughly before it is put in the aquarium. This is best done with lukewarm water. The rinsed substrate is poured into the aquarium and spread uniformly over the bottom.

If you do use gravel in your tanks, it is necessary to stir up the gravel on a regular basis. This is done so that uneaten food and debris cannot accumulate under the surface, which causes the water to foul and turn the gravel black and malodorous, thus wreaking havoc with the fish. Plants are not necessary for the raising of guppies. If you want them you can have them, but be prepared for the extra work.

WATER QUALITY

Water changes may be the most important topic in this entire book.

Back in the fifties, the only water that was added to the aquarium was replacement for the water lost to evaporation. It was found that the fish originally in the tank remained healthy after the tank had been set up for some time, but when any new fish were added, they usually ended up sick or dead. What was occurring

Four-month-old blue delta breeder.

Dark blue delta. The dorsal is showing a good color match to the caudal. The body looks somewhat bent, but it was just the angle he was swimming when I took the picture.

was the gradual build-up of concentrated salts and minerals that were lethal to the newly added fish. The answer to this problem was simply draining water off the bottom of the tank, rather than just replacing that which had evaporated. By frequent siphoning, one reduces the salt and mineral content to an acceptable level.

With the smaller tanks it may be necessary to change more water more often. There is not the room for error you would have with larger tanks. More attention to detail in feeding and filter management is also required in order to prevent

AQUARIUM MAINTENANCE

Water conditioners make it possible to use freshly drawn tap water for water changes. Photo courtesy of Hagen.

any runaway change in pH and ammonia levels. Along with the increased water changes there is less impact on your fish from both these factors. While closer observation and a little more work may be involved with the smaller tank set up, you can still do a good job with your fish.

As the number of my tanks increased, I stopped using gravel and went to bare tanks. The mulm and any uneaten foods were then more visible on the bottom of the tanks, so about once a week I siphoned this off and filled the tanks up with fresh water of the same temperature. After doing this for a while, I noticed the fish looked better and were eating more. This led to a lot of experimenting with the frequency and amount of water changed. With the infusion of fresh water, the fish really perk up. It must seem like a fresh rain to them.

I tried daily water changes of approximately 10 to 15%. The fish did very well in both growth and health, but for me it was just too much work and often I didn't have enough spare time to devote to this program. I reached sort of a compromise by changing roughly 25% of the water once a week. The fish didn't grow quite as fast and the tanks aren't quite as clean, but I am able to maintain this schedule. Whatever the type of water changing program you set up, it is necessary to adhere strictly to it. If you don't, the fish will suffer.

With young fish, you can change up to 75% of the water without too much harm. As the fish get older, you will have to cut down on the amount of water changed and also the frequency of the changes (especially on show-size males). I would say 10 to 15% once a week is plenty.

I do not find it necessary to add any chemicals to the replacement water. I take it

Commercial remedies are available at pet shops to combat a wide range of aquarium illnesses. Photo courtesy of Aquarium Products.

directly from the tap. This was true when I used chlorinated municipal water and it is true with my present well water—even though well water varies somewhat with the seasons and municipal water varies in the amount of chlorine present. Occasionally, in different parts of the country, water conditions are not conducive to keeping fish. In these instances it may be necessary to bring in water from another source. Normally, these problems can be controlled with filtration or softening of the water.

My own water supply is from a well on our property and is fairly hard. I run the hot water through a softener, but the cold water bypasses it. By mixing the two, I still get hard water, but not as hard as if there were no softener involved. It did take some conditioning of the fish to get them acclimated to the well water, but once I had babies born into my present water supply, half the battle was over.

Some regions of the country have excellent water for raising guppies, others do not. Especially poor are those where the municipal water system adds additional chemicals to the water. Those unfortunates will just have to watch the amount of water they change, especially in the summer. I consider it a waste of time and effort to try to maintain a constant pH level, at least for raising guppies. The fish must become acclimated to your water conditions rather than you trying to artificially maintain theirs.

It would be possible to devote many pages to the chemical makeup of the aquarium water, but unless you are planning to do a scientific paper on it, who cares? Just set up a workable system of water changing and adhere to it. For those of you setting up a tank for the first time: If no conditioned water or used filter are available, you

AQUARIUM MAINTENANCE

Red delta R X R line with very good dorsal to caudal color match. Note how the color flows well up into the body.

can add a stress-reducing product to your water. While not as good as fish-conditioned tank water, it is much better than just raw water.

Another advantage to changing the water on a regular basis is that it helps keep the ammonia and nitrate levels in the tank down to an acceptable level. As the hobby has progressed over the years, we have come to understand the importance of these two factors. But make no mistake about it, failure to perform water changes will also cause you to fail in your efforts to raise show-quality fish.

I do not use a storage tank to age water, nor do I add any conditioners to the water. Replacement water is taken directly from the tap via a garden hose. The only requirement I have of the water I use is that it be around 74°F.

When introducing new fish to your tank, it is important to know the pH factor of the water these fish were raised in. If it is more than a couple of tenths of a point different from your own water, you will have to gradually adjust or acclimate them to your water conditions. They seem to be able to accept a variance in the hardness factor, but too much of a difference in the pH level will cause them to fall apart.

I have found that it is possible to maintain the caudals on the older, larger males by placing them in a 2-quart or

Red veil bred by Jim Alderson. Good all 'round color match makes this a good example of the veiltail guppy.

AQUARIUM MAINTENANCE 59

Blue delta line, a medium blue with very wide tail and a good color match in the dorsal. The red body patches are not desirable in the blue class, but are very hard to eliminate entirely.

gallon-sized drum bowl. No filtration or aeration is used. Water for the bowls is taken from an established working tank; no fresh water is used. The bowls are cleaned every week and the water replaced. Light feedings of live baby brine shrimp, my dry food blend, and frozen adult brine shrimp are fed. Very little or no increase in size is noted, but the caudals hold up surprisingly well.

After having the males in these bowls for a period of time, I add a female or two. In the more confined space males of 12 to 14 months of age are able to breed, whereas in the larger tanks they have problems catching the females. You might want to try this technique if you run into breeding problems.

I try not to change the water on the males shortly before a show; if I do, the tails will invariably split on the fish.

It is at times difficult to maintain the clarity of the water in your tanks. I have seen fish people keep crystal clear tanks and raise poor fish, while at the same time some tanks that you

AOC bicolor delta. The caudal-dorsal match is fair, the pattern could be a little better, and the body is a little thin.

Albino delta with a good color match and nice long dorsal. Albinos are usually smaller than most show fish.

can only see an inch or two into produce excellent fish.

One time, while fooling with direct lighting for raising water sprite, the tanks that were planted turned a soupy green as result of excessive light. If the fish moved away from the glass they were not visible at all. Just for the heck of it, I left this setup going for a couple of months. There didn't seem to be much difference in the growth rate as compared to other fish of the same age group, but the color seemed to be somewhat more intense.

Don't be concerned if your water isn't perfectly clear; it is difficult to attempt to feed fish the amount of food that is necessary for proper growth and not have slight bacterial build-ups at times.

I have found it best to change the filters a couple of days before a water change, as this seems to maintain the balance better than the other way around.

Keep experimenting with the frequency and amounts of water changes until you find what works best for your fish.

If you *are* having problems with your fish and you suspect that the water is the culprit, you may have to go to storage tanks to condition your new water before adding it to your tanks. This is a lot of extra effort as they take up a lot of valuable space and you must use some kind of pump to transfer the water after it has aged.

Before going this route, I would try using carbon in a

Young red delta. Dorsal and caudal are a good color match. The body also has good coloration. This is one of my breeders.

AQUARIUM MAINTENANCE 63

Double sword bred by Don Saurers; nice fish with good body size and fair color. Both swords are equal in length with good shape. The dorsal has the ideal shape for a swordtail and only slightly off in color.

line filter on your water supply to your fish. Many times this will do the job in protecting your fish from wild swings in water quality. All too often people will lose many of their fish due to additional chemicals added to the water supply. While this may be okay for humans, it can be tough on fish.

The worst part of it is that when your water supply goes bad, you don't notice it until your fish start dying off *en masse*. Usually there is not much you can do for them when they reach this stage. This problem seems to be more prevalent on the East Coast.

I have talked to people who have had to haul water over 50 miles to use in their tanks to keep their fish going. About the only advice I could offer if they

AQUARIUM MAINTENANCE

had tried everything else, was to change much less water at a time, at least until the water supply had returned to normal. You have to admire them for sticking with their fish under such adverse conditions.

WATER CHANGES

There are many siphoning devices on the market to make your water changing chores easier. The only advice is that you find a system that you are comfortable with and keep up those water changes!

Half black red veil bred by Frank Orteca with good color, but the dorsal has some off-color spots. Body shows fairly good demarcation line with good black color.

AQUARIUM MAINTENANCE

Variegated snakeskin bred by Frank Orteca. Good dorsal and caudal match. The body color is also good, but the snake pattern is only fair.

Guppies are particularly contented and also grow better when a partial water change is carried out at least once a week. In a show tank the partial water change is somewhat more difficult under certain circumstances because the aquarium is located some distance from your water supply. This is no excuse for neglecting it, however, and you should get used to changing 10 to 20 percent of the aquarium water once a week. Food remains lying on the bottom are a problem for the water quality of an aquarium. In smaller show tanks, in particular, you must make sure that these food remains are siphoned off without fail.

In the fish room I use a length of rubber garden hose to siphon my top row of tanks. Most hardware stores will be able to supply you with whatever length

AQUARIUM MAINTENANCE

you need. Always buy a little more than you think you need. After a little practice you will even be able to get the siphon started without getting a mouthful of water. To drain my bottom row of tanks and those not near enough to the drain pipe, I use self-priming pump. It is mounted on a support board used as a base that keeps it about an inch or so off the floor. I attach a 12-foot length of garden hose to the inlet and one of the same length attached to the

Light purple out of my purple line. The fish was frightened when I took the picture and stood on his tail. He has good shape and color, but the dorsal is somewhat off in color.

AQUARIUM MAINTENANCE

Solid snakeskin bred by Frank Orteca. The caudal color and body pattern are good but the dorsal is small and mismatched.

outlet. The outlet hose is mounted by a clamp to the edge of my sink to make sure it doesn't come loose and run water all over the floor. On the outlet hose I have attached a 3/4-inch piece of PVC pipe about a foot in length with an elbow connected to an end-plug. I have drilled a series of small holes on the bottom section. This prevents fish being sucked into the drain. The pump is portable and with the use of a grounded extension cord, I am able to service all the tanks in my fishroom. All in all, the pump sure beats hauling buckets of water!

Be sure to stir up the gravel (if you are using it) before you do your water change. This is done so that uneaten food and debris cannot accumulate under the surface, which causes fouling of the water and wreaks havoc with the fish.

CLEANING THE TANKS

Cleaning the entire aquarium is quite simple. To break them down completely you must drain all the water out, remove the equipment and plants, and, of course, most importantly, the fish. However, the fish *are* easier to catch after any

Dark purple with super wide tail, excellent color, and a fairly good match of the dorsal.

AQUARIUM MAINTENANCE

Medium blue delta line. This is about as light a blue fish as I raise. He has good color in the body, dorsal, and caudal and a very good body for a young fish.

accessories are removed and if the water level has been reduced first, giving them no place to hide and less room to run.

After the aquarium is empty, clean with a strong salt solution in a gallon of warm water in the tank. Using a nylon sponge pad, wash the tank completely, paying particular attention to the corners. After the tank is clean, rinse several times with clean water before setting it back up. If you want something stronger than plain salt as a disinfectant, diluted liquid chlorine bleach is very effective. Just be sure you rinse the tank thoroughly after use. When you refill the tank, use a chlorine remover as well to be sure no trace of chlorine remains to damage your fish.

CLEANING THE FILTERS

Cleaning the filters is relatively simple. I use a 5-gallon plastic pail filled about 3/4 full of cool water, and add a cup or two of liquid chlorine bleach. After removing the filter floss and rinsing out the filters, they are placed in this solution overnight. Remove the filters and thoroughly rinse off with cool water. Remember to unplug all the air holes. Marbles and floss are then replaced in the clean filter and the filter is ready to be returned to the tank. Using this method, I can usually clean up to a dozen filters at a time with very little work involved. When setting up a new tank, I fill it 2/3 full of water drained from other working tanks. A used filter is added to the newly cleaned tank and run for a

A red veil from the line showing good color and finnage. This fish is about 5 1/2 months old and will reach full show size in a couple of more months.

Half black red delta displaying a well-shaped dorsal and caudal. The fish's color did not look good with the type of lighting I used for this shot. This is a problem we have at some of the guppy shows.

while. Then fresh water is added to finish filling up the tank. Usually I let the filter run for 24 hours before adding any fish, but if necessary, they can be added right away. The main thing is that you do not want to shock the fish by hitting them with fresh water and a new filter at the same time.

There are no hard and fast rules about how frequently filters must be cleaned. Under conditions where you are not feeding excessive amounts of live baby brine shrimp, the filters can be left for a month or even several months with no ill effects. This applies more to tanks on the bottom racks with lower temperatures. The bacterial balance is more easily maintained under these conditions.

In tanks where you are feeding heavily with live brine shrimp as well as large amounts of dry food, it may be necessary to clean your filters on a weekly or bi-weekly basis.

If your filtration system is working properly there will be very little mulm on the bottom of the tank. I realize it is perhaps difficult to envision a balanced aquarium without gravel, plants, etc., but if you can maintain the tank without overfeeding and the filter has the proper bacterial balance, the water will remain clear even though the fish are somewhat overcrowded. Fish raised under these conditions remain healthy and grow well. I just wish I could keep all of my tanks this way all the time!

On the other hand, if you

Group from a red delta line showing good caudal and dorsal match. They should be ready to show in about a month.

A blue delta from my line. He has a good caudal except for a small nip in the top portion. The dorsal matches in color but was slightly folded when the picture was taken. He has a good body shape with fair color.

do notice an accumulation of mulm on the bottom of the tank, primarily from uneaten food, you will also notice that the filter is starting to plug up. At this point I would clean out the tank and set it back up. You could, of course, suspend feeding, but by this time it's probably too late.

OXYGEN CONTENT

At one time I became interested in the oxygen content of the water with regard to the growth rate of the fish. I reasoned that a higher oxygen saturation would induce the fish to be more active, less disease prone, eat more, and thus grow more quickly. As most of the oxygenation takes place on the surface of the water, I then reasoned that if I had some tanks with a larger surface area, it would be better for the growth process. With this in mind, I purchased a number of tanks that were 24 inches wide and 30 inches long and only 8 inches deep. These tanks were setup, and I anticipated growing the "super fish." After about six months there are no noticeable difference in the growth rate as compared to fish raised in the 15-gallon tanks. Next, I thought to raise the temperature, but that would be self-defeating, as there would be even less oxygen saturation. From there, I went to increasing the air flow to the filter in order to bring more carbon dioxide and other gases to the surface for more effective oxygen exchange. Still, the volume of the water and the

Pingu® male showing the unusual body color of this type of guppy. Not much in the way of finnage or color and usually they do not do well at shows. Bred by Don Saurers.

Half black red delta with large dorsal.

temperature determines the amount of oxygen saturation. Also, increasing the air flow created too much turbulence, making it difficult for the fish to feed or even swim in a normal manner. A short time later those tanks were disposed of, and I went back to the 10- and 15-gallon tanks.

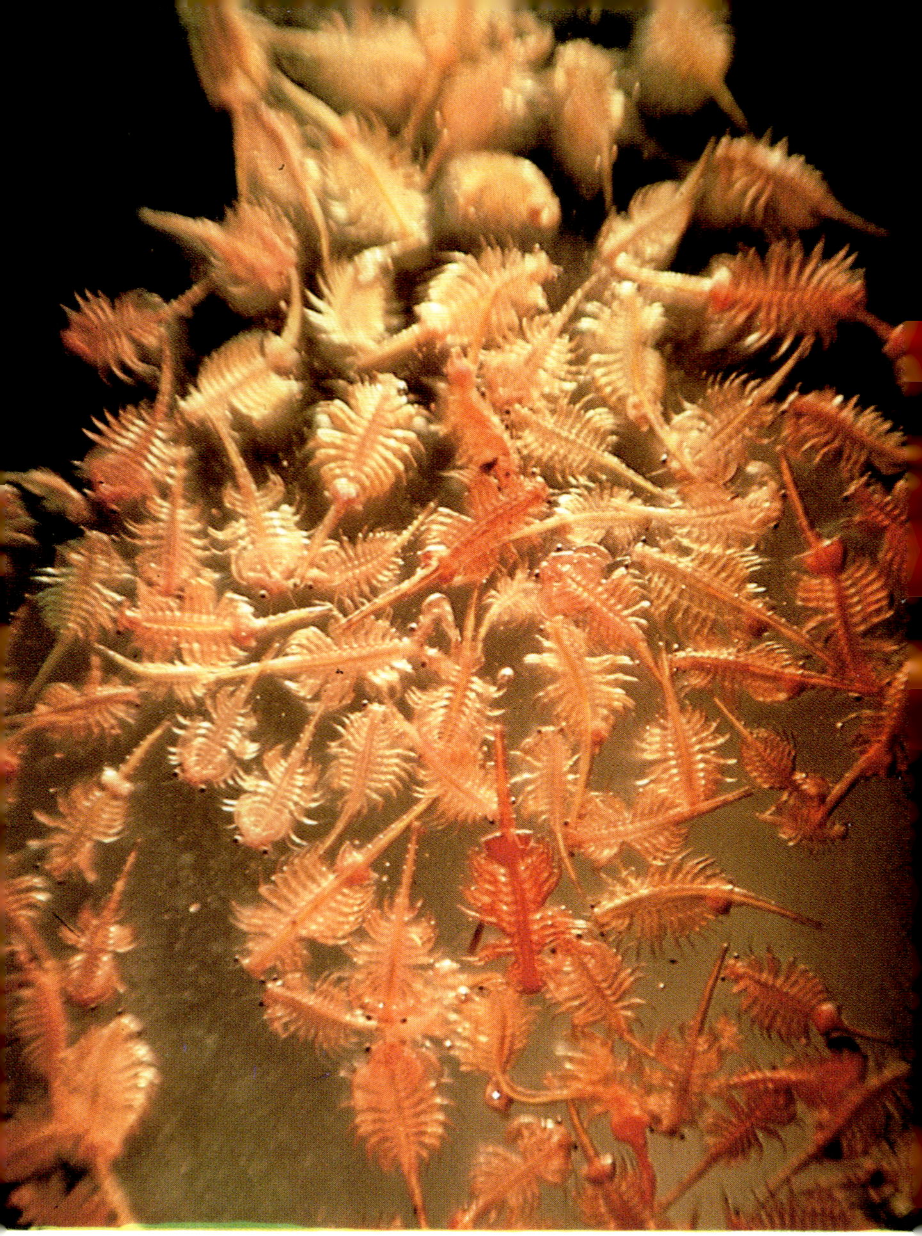

Feeding Your Guppies

Probably the most important factor concerning food for guppies is "Will they eat it?" I realize that sounds somewhat simplistic, since if you were to starve them for a long period of time, they would probably eat ground-up newspaper, but on a day-to-day basis, it is necessary to provide them with a palatable food that will entice them to eat more, even with partially full stomachs. An active guppy can probably eat every hour or so. But if what they eat appeals to their taste buds, they will continue to feed even though they are not really hungry. Just like trying to eat just one potato chip.

BRINE SHRIMP HATCHERS

The type of brine shrimp hatchers I use are one-gallon jugs with the bottoms removed. I happen to use glass jugs, but other people have had good results with plastic soft-drink bottles as well. The easiest method I have found to remove the bottoms is as follows: clean the outside of the jar with warm soap and water. Rinse and dry with a towel. Lay the jug sideways on your lap at a slight angle, place a glass cutter in the appropriate hand and apply to the jug, slowly revolving the jug with the other hand until you have

Facing page: *Brine shrimp are eagerly taken by guppies. If a fish won't take live brine shrimp you know it's very sick! Photo courtesy of San Francisco Bay Brand, Inc.*

Some people use 2-liter soft drink bottles to hatch their brine shrimp eggs. This system uses three bottles in rotation to ensure a constant supply.

made a cut ring around the bottom of the jug. Next, fill a basin or sink with cold water approximately three inches deep. Handling the jug carefully, fill it with about an inch of hot water. Holding it by the neck, dip the jug into the cold water. The bottom will fall right off. Round off the sharp edges with a file or emery cloth. Now the jug can be cleaned thoroughly on the inside. At the hardware store, you should be able to pick up a cork to fit the neck of the jug. Drill a hole approximately ⅛ inch or slightly larger through the center of the cork to accommodate a length of flexible airline tubing. You should have a rack built to

FEEDING YOUR GUPPIES

support at least three inverted jugs. (As an added precaution, it's a good idea to run a strip of plastic filament tape around the rim of the bottle in case you bump it against something. This helps to prevent breakage.) The air control valves are placed above the jugs and the tubing from each jug is then connected to them. The light is placed near the center of the cluster of jugs so that the necks of the jugs are directed towards the light. When the air is shut off, the live brine shrimp will be attracted to the light. After they settle for a few minutes, they can be siphoned out with a meat baster and fed to the young fish. By using four jugs, it is possible to provide a 48-hour hatching period and still feed live brine shrimp every 12 hours.

The hatching solution consists of ½ cup solar salt (available where water softeners are sold), ½ tablespoon of epsom salts, and one teaspoon of baking soda. To this is added about three quarts of slightly warm water. It may be

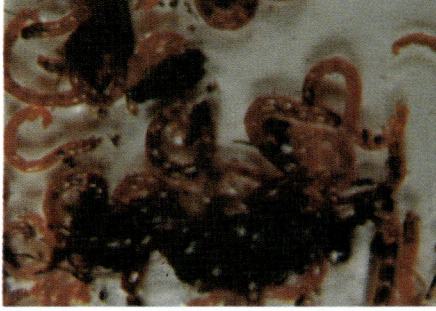

Bloodworms are not worms at all but the larvae of a midge. Adult guppies are very partial to this food but it is a bit large for fry.

necessary to vary the hatching solution slightly to determine which blend gives you the best results with your water conditions. After feeding your fish, thoroughly wash out the baster with fresh water; this will help prevent disease problems. If you use a hydrometer, the reading will be from 1.011 to 1.025. A visual indication of excessive salt is when a large number of brine shrimp are found around the top of the jar when the air is shut off. I have found that when using Great Salt Lake eggs, they require a lot more air in the hatcher for a good return. About a week's supply of eggs are kept in the fishroom in a tightly sealed container. All the rest are stored in the freezer. If you note that shrimp hatches seem to be falling off, place the eggs in a 250°F oven for 15 minutes. This seems to drive out some of the moisture, producing improved hatches.

The hatchers may be run up to 48 hours. After this time, the shrimp will start dying off. Aerate to where the eggs and live shrimp are

A tremendously enlarged photo of brine shrimp a few moments after hatching.

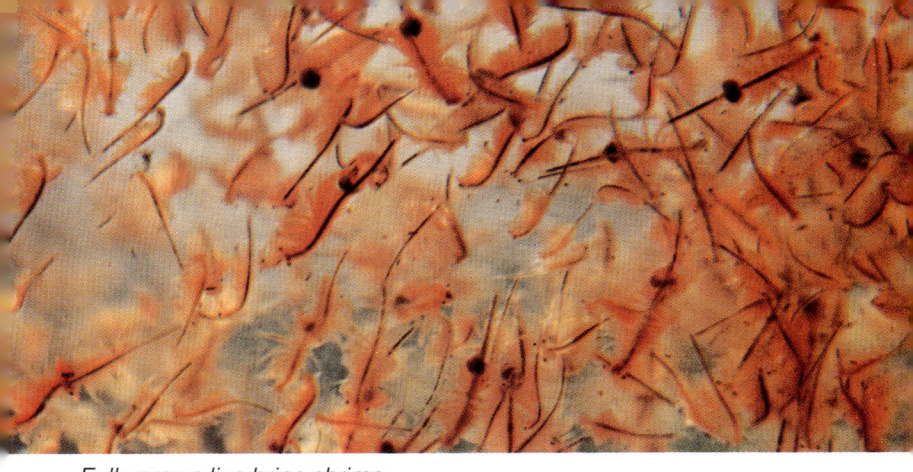

Fully grown live brine shrimp.

moving well in the hatcher, but not to the point where it damages the shrimp.

Through the use of four hatching jars, you can get optimum hatches. However, the leeway time between a good hatch and dead, overhatched shrimp is only an hour or two, so for this reason, I have cut back to three hatching jars. This gives you a hatching time of around 36 hours per jar. This way, if you are delayed in feeding your fish, you will still have a pretty good hatch. I would not recommend feeding dead, overhatched shrimp as they turn foul very quickly and can really mess up your tank.

If too much aeration is used in the shrimp hatcher, the eggs will stick to the side of the container and will not hatch properly. There are various grades of shrimp eggs, each requiring different hatching solutions and hatching times as well as temperature levels. Your water chemistry will have a great deal to do with what your hatch rate is. Most of the time I use a commercial grade of egg which gives me a very good hatch rate. I have tried eggs from Utah, California, Canada, China, and South America and mainly stay with the Utah eggs as most cost effective.

If you have difficulty maintaining the 80°F temperature in your hatchers, you can build a cabinet to keep the heat around the hatchers. Another way is to put an enclosed aquarium thermostat without the heating element in series with your light bulb to keep a constant level. In my fishroom, the temperature stays about the same all the time, so I am able to use the same wattage bulb all year round.

FEEDING THE FRY

So let's start with feeding the baby fish. Live baby brine shrimp is the ideal baby food. This is fed twice daily until the fish are four months of age. At this time I discontinue feeding brine shrimp to all, except for pregnant females. The reason for this is that when the fish are younger they usually clean up all the shrimp. As the fish mature, quite often a certain amount is left uneaten. These die and have a tendency to foul the water.

FEEDING SCHEDULE

For the first feeding of the day, I shut off the shrimp hatcher which has competed the 36- or 48-hour cycle. The fish are then fed a light feeding of my dry food. After a couple of minutes I go back around and give those who have finished up the first feeding another light feeding. Occasionally I will go around again a third time.

After about 15 to 20

minutes, the live shrimp have settled in the jar and they are siphoned out and fed to the young fish, usually enough for them to feed on for several hours. Another feeding of dry food is given to the others who did not receive any live shrimp.

When I get home from work at night, they are fed a couple of light feedings of dry food. About three times a week, all of the fish, except the very young babies, are fed adult frozen brine shrimp. This is broken off from a pack and placed in a cup until it thaws. Then it is placed in a net, rinsed several times with cool water, and added to the tanks. Feed them all they will eat without leaving any leftovers.

A couple of hours before the lights go out, I again feed the baby brine shrimp in the same manner as in

Adult brine shrimp will vary in color according to their diet. Brine shrimp feed on algae in nature but can be maintained in captivity on a diet of dissolved yeast.

the morning. That is my total feeding schedule. If you have the time and inclination you could feed your fish 12 to 15 times daily. Let's face it—the more they eat, the more they grow. Basically, it comes down to that point. With a better quality dry food it tends to make your job a little bit easier. It is hard to say how big my fish

Prepared foods come in many forms, from the finest powder to bite-size pellets.

would grow if I fed them as many times daily as some of the other breeders do.

FOODS FOR GUPPIES

Over the past thirty years, I have fed my fish everything from Wheaties® to whale meat and about everything in between. Some of the commercial guppy foods are pretty good and some are just better than nothing at all. There are a lot of foods that would be good for your fish, but because the fish are cold-blooded (their body temperature is the same as the water), these foods will not break down and thus will pass through the digestive system intact and so you are forced to use food groups that can be assimilated. As to whether it is to your benefit to blend your own food mixtures or to use ready-made foods is up to the individual. I make up my own, as I cannot buy prepared foods with all of the ingredients I wish to use. It is a general practice for fish food manufacturers to put the better quality ingredients in the higher priced foods, so you might use this as a guide when purchasing foods. Don't try and save a few cents by using low quality foods as they will cost more in the end.

Is there one dry food that has all the necessary

FEEDING YOUR GUPPIES

nutrients and vitamins for a completely balanced diet? The answer would have to be "no," thus the reasoning behind the blending of my own food. It consists of various fish and fish liver meals, shrimp and other crustaceans, beef liver and selected meat meals, kelp, some cereal meals, spirulina, and various yeasts along with many types of vitamins and minerals. I realized a long time ago that it was impossible to duplicate the various foods fish would eat in its natural environment. So why try to act like Mother Nature?

I used to feed a dozen different foods, one at a time. I didn't always get the results I wanted using this method, so I decided to blend them all together, and now I am getting much better results.

Usually when I mix up

Blackworms (upper left) are a newer form of tubificid worm than regular tubifex (lower) and both are safe when well rinsed before feeding.

my food I have to close the fishroom door, otherwise I have my wife, Ethel, and the cat (of course) visiting me; my wife coming to say that I'm smelling up the house, and the cat coming

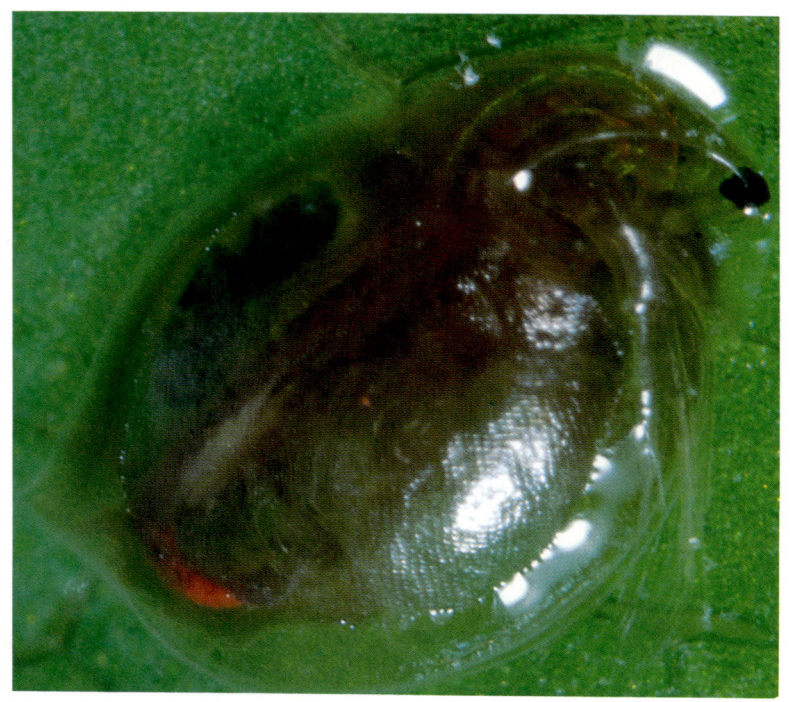

Daphnia, a freshwater crustacean that is a favorite live food for fishes.

to see what smells so good. After all the mixing and cleaning is complete, the food is placed in a double plastic bag, sealed, and then put into the deep freeze. This keeps it fresh and helps prevent the food and vitamins from deteriorating. About a two-week supply of food is put into a sealed glass jar and kept in the fishroom for my use.

When feeding dry foods, use a fine or medium-fine grade, as the particles must

FEEDING YOUR GUPPIES

Tubifex worms tend to clump together. This is an advantage at feeding time since the worms are easy for the fish to find.

be small enough for the fish to ingest. If using flake food, it's a good idea to crush it between your fingers for the baby fish. One point I should make is to have your hands clean when feeding your fish. You could introduce some harmful substance and then wonder why your fish are sick.

A few years back, while feeding my fish, I decided to experiment with a tank of young females. I began by dropping a slight amount of dry food in the tank, only enough to partially satisfy their appetites. A few minutes later I did the same thing. I continued doing this for about 30 minutes. During this time the fish became so worked up, they literally boiled the water into a froth trying to get at the food. Even though their stomachs became greatly distended, they still wanted to eat more. I stopped the experiment in feeding frenzies when several of them dropped to the bottom of the tank—dead. So what did it prove?

FEEDING YOUR GUPPIES

Perhaps only that the way you feed, combined with the quality of the food, determines the growth pattern. It would be nice if it were possible to retain this triggered response in all your fish at feeding time, but it doesn't work that way. Without belaboring the point, any time the fish do not come to the front of the tank to feed, do not give them any food, no matter what you may think they need. Even if it becomes necessary for them to go a day or two without any feeding, they will be healthier and better off for it.

The type of food you feed may also determine, to an extent, how many young the female drops, as well as have a bearing on the male to female ratio. This is why it is important to have a well-balanced food mixture.

As you have probably noticed, I do not use any other live food except baby brine shrimp. Initially, I had thought it was necessary to feed additional live foods to complement the dry foods for good growth. But in my own case, all the extra work in obtaining live food, along with the possibility of introducing disease to my tanks, made it unattractive.

Tubifex worms are probably one of the best growth foods you can use, but caution must be utilized in handling and feeding. They require constant flushing and must be kept with water running over them or in the refrigerator. I think I, and certainly my wife, would draw the line at keeping worms in there. I have seen some exceptionally nice fish raised using tubifex worms, but also have seen a number of guppy raisers

Above: The life cycle of the mosquito. Guppies love mosquito larvae. **Below:** Midge larvae, or bloodworms, live in little tunnels in the substrate. That's why bloodworms are sold frozen. They can't really be kept by aquarists.

put out of business because of them.

White worms are useful for body growth, but shouldn't be fed more than twice weekly.

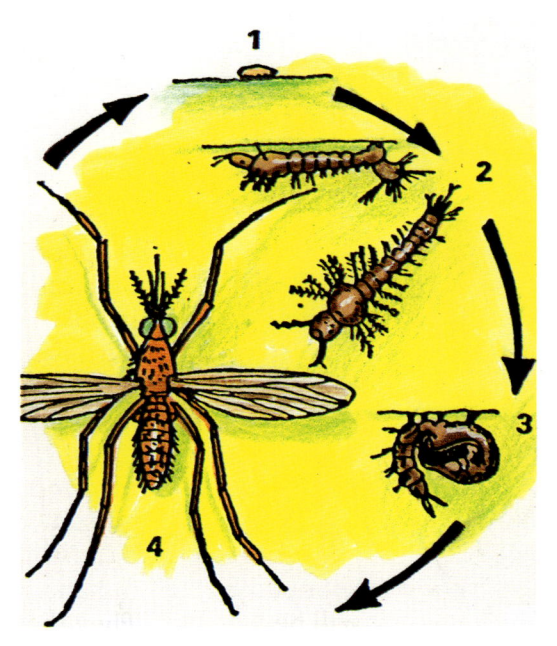

Overfeeding can cause a buildup of fatty deposits in the body, resulting in an extremely bloated condition and ruined fish.

Daphnia is also a fair fish food. The protein content is quite low, but the fish seem to enjoy chasing them and the roughage is beneficial to the fish.

FEEDING YOUR GUPPIES

Prepared fry food is handy to use and good for supplemental feedings.
Photo courtesy of Hagen.

Mosquito larva is also a very good fish food that my fish enjoy very much. The one and only time I used them in my fishroom, the fish didn't eat all of them and I ended up being chased by and chasing adult mosquitoes for several days.

If I don't seem all that enthusiastic about using the above-mentioned live foods, it is because I have found it unnecessary to use them. If you want to give them a try, by all means do so.

Another type of food you may want to try is a paste food such as a modified Gordon's Formula. This consists of beef liver which is deveined and put through a blender, then water is added as required. Next, some dry baby food is combined with powdered gelatin or agar-agar along with some water soluble vitamins. All this is mixed together and placed in a double boiler and cooked for a while. After cooking, let is cool down. Then place small amounts into plastic bags. Once they are sealed, flatten them down and place them in the freezer. When you are ready to feed your fish, just break off a chunk. This makes for a

Feed flake food sparingly until you can gauge how much your fish will eat.

pretty fair food, but if overfed it has a tendency to foul the water very easily. Beef heart may be used in place of the liver, but be sure to remove all the fat and veins. I have also tried chicken but the results were not as good as with the beef.

For a while I used to make up what I called my Sea Mix. It contained the following ingredients blended together: partially cooked and frozen salmon, tuna (water packed), shrimp, clams, oysters, eggs, kelp, salt, high protein dry food, and vitamins. It was very good but expensive to use, and now I get about the same results without it.

One thing about cooking foods is that you destroy

much of the vitamin content in doing so. It is advisable to keep this in mind when making up your foods.

There are two types of frozen brine shrimp available, baby and adult. The baby brine shrimp come in very small plastic packets or containers and are not really much use to a guppy breeder with any number of tanks. Adult frozen brine shrimp is an excellent food provided it was frozen when the shrimp were still fresh. A good way to check on the quality is to look at the color. If it is dark red it is usually good; however, if it looks black, chances are that it has thawed once and been refrozen. If this seems to be the case, do not feed it to your fish, as it can really make a mess of your tanks. There may be some variance in the color of the frozen shrimp, as some may have a diet of algae and as such may be greenish in color; these are okay, but if black, no. This also holds true with frozen daphnia or other types of frozen foods.

Tubifex worms in a freeze-dried state can be used with few or no problems. They usually come in little cubes and can be pressed to the side of the aquarium for the fish to nibble on. Females seem to really go for them, whereas the males don't much care one way or another. If using them, keep the unused portion in a well-sealed container; they seem to deteriorate quickly if any moisture gets to them.

It should be noted that you should never feed more food if there is any uneaten food on the bottom of the tank. Make the fish clean it up before they are fed again.

One advantage to feeding

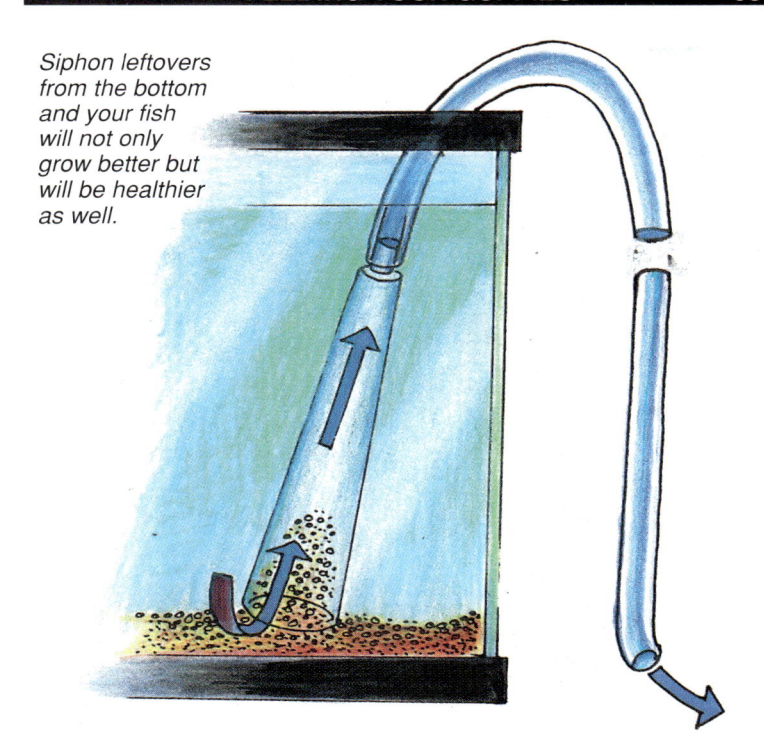

Siphon leftovers from the bottom and your fish will not only grow better but will be healthier as well.

any type of live food is that it seems to bring out aggressive behavior in the fish, causing them to eat more than they would normally. This is one of the reasons I feed dry food first, and then fed the baby brine to the young fish. The idea is to get them to think the dry food is all they are going to get, so once the live shrimp are added they will really stuff themselves.

I have had some bad results with some of the

commercial dry foods from time to time, but I do not feel that it would be proper to name the brands, as it is possible that I just got hold of some bad batches. But you sure can mess up a lot of good fish very quickly with bad food. The trouble is that by the time you catch it, it's usually too late, and months of work have gone down the drain.

All things considered, buy the best food you can afford, feed lightly as often as possible, and be very careful not to overfeed. Leave the lights on at least an hour after your last feeding to give the fish a chance to clean it up. Try to maintain a regular feeding schedule. If all these rules are followed, your fish will grow well.

If you find it necessary to be gone a couple of days, don't worry about feeding your fish during that period. A day or so without food will do no harm and your tanks will be nice and clean when you return, to say nothing about how glad your fish will be to see you. The only ones you may have trouble with are the females that are about to drop. The others will probably be much better off than if you tried to have some inexperienced person come in to feed them. Only over extended periods of time will it be necessary to have someone feed your fish. I would keep it as simple as possible with very light feedings once or twice a day. That way you shouldn't have a mess when you return. Better a little hungry than to have diseased tanks.

Probably the two most important things to check when feeding your fish is if they come to the front of the tank to eat, and if the

A varied diet will bring out the best colors in your fish.

bottom of the tank is cleaned from the previous meal. By observing just these two simple feeding rules, and by not adding any additional food unless they are met, you will measurably increase your success in raising good, healthy fish.

Another thing to keep in mind is if you have a self-defrosting refrigerator, keep any frozen shrimp in a well-sealed plastic bag. If the container is open slightly, the refrigerator will draw the moisture out of the shrimp, leaving a semi freeze-dried food. I had deliberately tried this with shrimp and some other foods, but found that the fish didn't seem to care for the taste. Also, unless your dry foods are in a glass or sealed plastic container, they will rapidly go bad in the moisture-laden fishroom, so these should be kept in the freezer or refrigerator.

Years ago I used to buy various dry foods in fairly large amounts and didn't always keep them in the freezer. When I would finally get around to using them, many were spoiled or had worms in them.

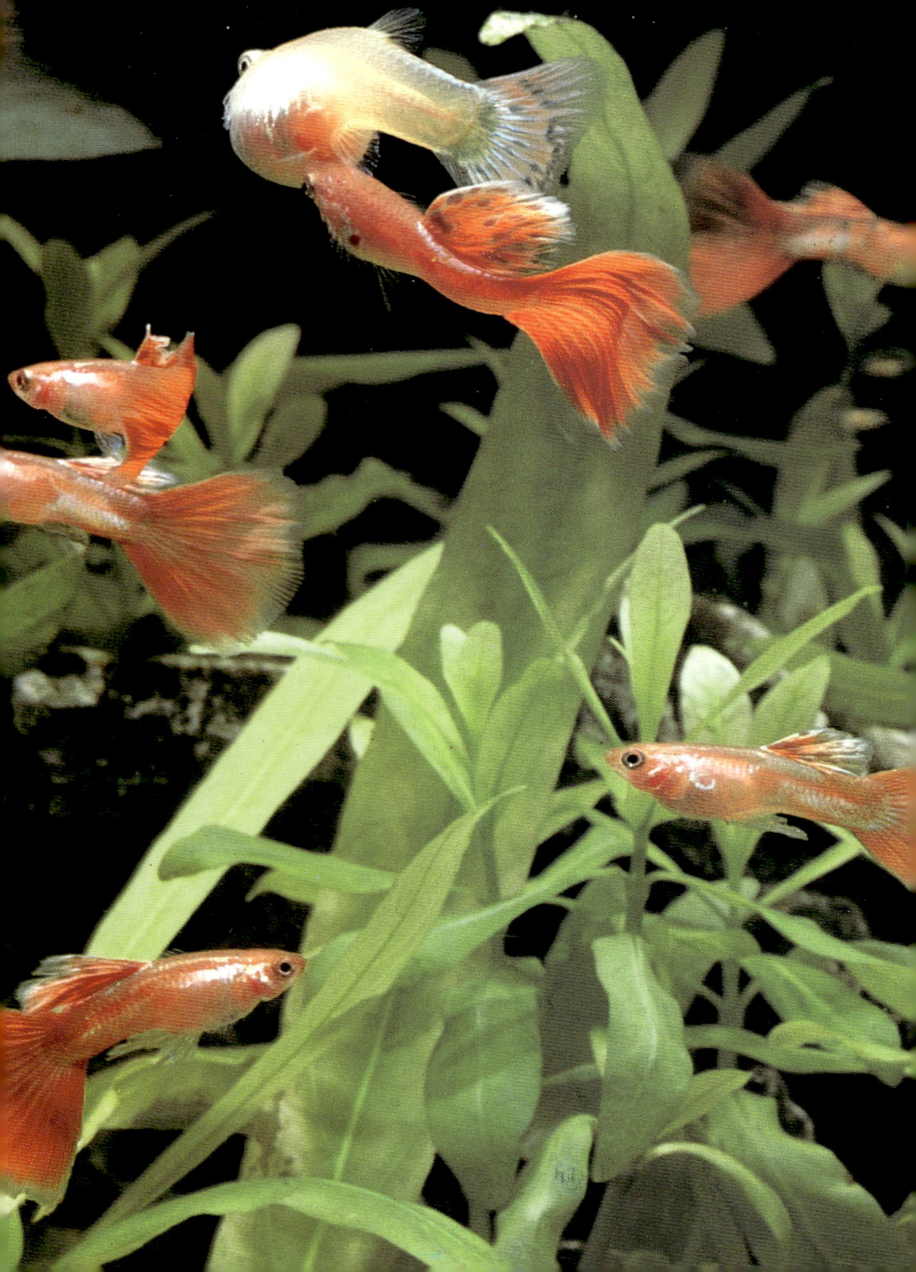

Breeding Your Guppies

Back in the fifties, when I started with guppies, information on breeding was rather vague or misleading. I read everything available and asked different fish people questions that they couldn't, or wouldn't answer. So I just figured to heck with it and started going my own way. I wasn't concerned that by crossing gray and gold guppies you would get a certain percentage of each type from the cross. My only thought was to

Facing page: Male guppies will pursue the females even when they are gravid. In a tank of mixed guppies, you will have fry of mixed parentage! *Below:* Half black red female showing a crescent pattern in her large caudal. She has a good-sized dorsal, though slightly off color; good body color. A good candidate!

increase the caudal size and improve the color.

My initial fish were from a couple of fish stores in the Detroit area; what you would probably now classify as semi-fancy guppies. Occasionally, a veil or a modified wide-tailed male would crop up. After fooling around for about a year, I started inbreeding the best of these fish. Within a few generations, the males started to look quite similar, with the caudals becoming more uniform in shape. However, there was still quite a variation in the caudal color. The basic color was a red variegated pattern. Occasionally I would get an off-color male with greenish, blue, or even black variations. After about seven generations, a gold male popped up in a dropping. He was bred to his sister, and then to his daughter, and finally to his grand-daughter. This was how I fixed my gold strain. About the same time, I took a couple of the off-colored males and started breeding them. The best way I found was to cross the son back to the mother and then to one of the best females of the resulting drop.

In this manner, I was able to fix, or set, the reds, blues, greens, blacks, variegated, and partial half-black strain. My original red variegated line was inbred for approximately 21 generations. Then, by using the females primarily for crossing the other lines, I was able to develop males with very wide tails, with some of the angles over 100 degrees, way beyond the

Facing page: *Females should not cannibalize their young. If your females consistently eat their young, eliminate them from your breeding program.*

BREEDING YOUR GUPPIES

99

super delta requirements. It would have been possible to continue breeding these super wide-tailed fish, but they were way beyond the delta standards of 60 degrees, so I stopped using this type of fish for breeders.

The reason for this is the possibility of one of the males being sterile. In this case, you could sit there for a couple of months watching the females fill up with unfertilized eggs and dropping no young. Most of the time they will just cast

As the gene pool narrows, it is possible to almost match the female's caudal color with that of the male. This is done without the use of hormones or color foods. One of my young half black red breeders.

BREEDING TECHNIQUES

My normal breeding setup consists of two males and three to four females. these out, but I have seen them absorb the eggs also. Only if a male is truly exceptional will I use just a single male, and then I will

Above: Half black AOC female with very good caudal development and color pattern. The dorsal also matches.

Below: After dropping her young, this female shows no interest in them. Some guppy lines are very cannibalistic and will eat all of their young if they can get to them. By selection, I have been able to pretty much eliminate this problem.

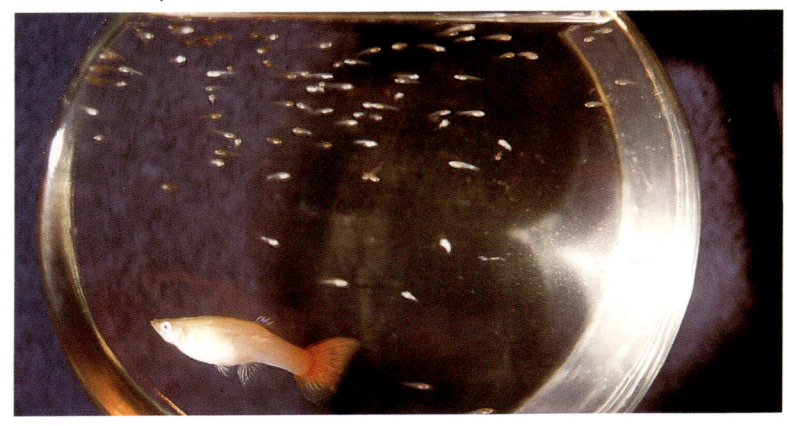

watch closely to see that young are forthcoming.

When purchasing stock I suggest getting at least two trios of fish, two males and four females. I would ask the breeder to set the fish up at least a week before you take them, especially if the females are virgins and all you are able to buy is one trio.

The fish are placed in a 5- or 10-gallon tank. Usually, I do not pick breeders until they are at least four to five months of age, the reason being that with some of the males, the caudals develop extending lobes as they get older. So I wait until it can be determined that the caudals are developing fairly evenly along the trailing edges. Basically it is a good idea to breed for one thing at a time—by this I mean that if you are breeding for size, use the largest fish; if for color, use the best colored males.

In a way I am fortunate

This female is a cross between gray-bodied and gold-bodied reds. She dropped about 60 babies a couple of days after this picutre was taken.

This Best of Show female was raised with several of her sisters in a small tank with no males. She was over three inches long and usually I am able to get them to breed with little trouble even at this size.

in that my lines have been around for a long time and are quite fixed. This enables me to use the best all-round males for my breeders.

The Breeding Stock

It is quite easy to look at a tank of males and pick out the best two or three fish. By best, I mean the ones who fill the requirements of the type of fish you are trying to develop. Remove these males and place them in a breeding tank. Normally, I leave them alone for 24 hours to establish their territories. For guppies, this means for them to simply feel at home in their new tank. The next day, the females are placed into the tank. This procedure is followed for all setups. If the females are introduced before the males, they pick on the males, often damaging their tails. It seems that the first fish put in the tank has pecking precedence over those put in later.

Picking the females is very important. I will sometimes spend quite a bit of time in choosing the correct ones for breeding. It is here that many breeders run into trouble in that they simply do not know what characteristics to look for in a female guppy. If this is not done properly, all you can do after a couple of generations is go back and get some more breeding stock. A lot of breeders cross everything and occasionally come up with some good fish, but again they are unable to hold the line.

I know of one top breeder who would place a bunch of different colored males and females in a 50-gallon tank for breeding. When ready to drop, the females were placed into a separate

A female from my gray-bodied half black line. Excellent body with a good caudal. This is the type of female I pick for a breeder.

A blue breeder female about ready to drop. She will be transferred to a 2-quart drum bowl to have her young.

tank. When they dropped their young, if they turned out red, they were a red strain. If by chance they turned out green, they were called a green strain and so forth. He did produce some really nice fish, but what a mess trying to maintain a true strain.

Each strain may vary somewhat, but generally speaking what I look for are females with a compact rather than an elongated body—the peduncle area of a good thickness—and a generally well-shaped body. (If they are all about the same, the nod would go to those who are more active and hungry.) Next, check the tails, some color is desired—usually that which is evenly distributed. Try to pick those with the widest separation between the rays, and yet at the same time maintaining an even tail shape. (By this I mean not having a large chunk or section missing.)

Most of the good females have a delta or wide box shaped tail. Occasionally, good shark-tailed females will throw the best deltas. When working with a cross, I will usually use one of each type until it is determined which throws the best males. These will be used for future breeding.

For those raising females for show purposes, it will be found that the gaudy or brightly colored females will not as a rule throw very good males. In fact, some of the nicest finned and best-colored females I've seen throw nothing but junk males.

It is possible to color test females using testosterone, but it is really not necessary any longer, as in most cases it is possible to buy almost any color strain you might want.

About the only time you would use clear-tailed females for breeding is with swordtails. When working with delta-tailed strains, you get very, very few clear-tailed females. Occasionally, when working with half-black reds, you will get regular gray females; these are good to use along with the half-black females for breeding.

Probably the hardest lines to maintain are the albinos. Next most difficult are a good line of reds. In the case of the albinos, it is necessary to cross the males to regular gray females and their daughters, otherwise you seem to run unto a sterility problem. Some lines you can inbreed for a while, but usually you have to revert to outcrossing. Most of the crosses are best accomplished with a green or red female.

One of my gray-bodied half black red females. The pattern is reminescent of red fish raised by Bill Sternke years ago.

To maintain a red line, once you have eliminated all the color impurities, it usually requires at least two, and sometimes three, separate linebred or inbred strains of reds. Even with three lines you must be very careful in selecting your breeders; any hint of black or gray in the finnage and you are in trouble. Some say it is possible to clarify the reds by outcrossing to black or blue females. My own experiences were negative in this respect, but perhaps the females were not compatible.

Under wild conditions, a large fin development would be a hindrance for obtaining food, escaping enemies, and reproducing. But in the controlled environment of our fish rooms we eliminate most of these factors. However, at the same time we also encourage, through the elimination of enemies and by containing the females to a small area, breeding by inferior males. In natural conditions, breeding would only be done by the most vigorous and hardy males, thus carrying on the vitality of the species.

In the artificial environment we provide, it is necessary to eliminate the possible breeding by inferior males if we are to improve the lines so it becomes necessary to cull the fish that do not come up to desired standards. It may seem cruel to some, but if left unchecked, the strain would deteriorate very rapidly. Being realistic, there is no sense in wasting time, food, and space on any inferior fish. It's not like the cute little runt of a litter of pups who turns out to be the nicest one of the bunch. All a runt fish can do is cause you trouble. This would apply to those of you who do not separate the males and females at an early age. Because, as sure as God made little green apples, the least desirable male will be the one that hits the female. Thus, all your efforts would be hindered and possibly doomed from the start.

I personally feel it is of prime importance to keep virgin females of each strain available for breeding. While there are some who may not agree with me, I can only say that it has enabled me to continue to raise high quality fish while others are having difficulties.

BREEDING METHODS

We have basically three different breeding methods with slight variations of each. They are inbreeding, linebreeding, and outcrossing.

A blue female, a good match for a splendid blue male.

Inbreeding

The breeding of closely related fish, such as brother to sister, father to daughter, or in-backcrossing, son to mother. The advantages of this method are in fixing a fin conformation, or a particular color or color pattern. Theoretically, you would take the best male and female of a dropping and breed them. Of the resulting young, you would do the same thing and so on down through the various

110 BREEDING YOUR GUPPIES

drops and generations. If you develop an exceptional male, you would breed him to his sister and if possible, back to his mother and daughter. This would give you a higher percentage of closely related fish for possible development of an outstanding inbred line. This practice may be continued for a number of years with only a slight diminishing of the quality of the fish, provided care is taken in the selection of the individual breeders. Usually you will end up with fish that look alike in both pattern and color.

In most cases, a female will have several drops of young once she is impregnated by the original male. If a different male is introduced to the female

Guppy embryos about eight days old. The eyes are very well developed at this point.

Fully mature females become so swollen when they are carrying fry that they look as if they could burst.

and provided his is viable, some of the next generation of young should be by him. This however, is not a hard-and-fast-rule. I have seen breeders try to use this method, and they only mess up their lines. They assume that females taken from a mixed tank and placed with the male of their choice will result in young from this male, but if the male is sterile you are stuck with the young from whichever male fertilized the female in the first place. In my estimation, it's not a good method for controlled breeding or maintenance of lines. While on the subject of inbreeding, if, for instance you want to breed gold or albino guppies, it is usually desirable to cross a male into a gray female. As gray is dominant over the recessive gold or albino color, all of the F1 dropping will be gray in body color. If

The circulatory system, including the heart, can be seen in these well-developed embryos.

The newly born fry straightens out quickly and immediately swims away to seek shelter and safety.

brother and sister are then bred together, the F2 young will be approximately 25% gold and 75% gray (50% gold-gray and 25% gray in genetic makeup). As you are trying to establish a gold line, you would use fish for breeding that display the gold color that you are looking for. In breeding brother and sister again with the gold-colored bodies, you should get 100% gold or albino, whichever the case may be. From then on, it is simply a matter of improving finnage and coloration. One of the problems encountered in working with the gold guppies is that while attempting to increase the body color, you have a tendency to lose the basic gold body coloration. To help prevent this, you would use only the males that display at last 50% of the basic gold body color. Even though you may not get as many body color points in the show, at least your fish won't be disqualified. I raised golds for a number of years, taking several class championships, and mainly dropped them because of lack of competition and the desire to use the tanks for other lines.

Linebreeding

Linebreeding is perhaps the safest and most efficient method of guppy breeding. In this, as well as in any breeding program, it is important that good records be kept, either in your head or written down. I would suggest the latter. To start off a linebreeding program you would take a male and breed him to two of his sisters. The resulting young

It is sometimes surprising how large a newborn guppy fry is. Baby guppies are completely self-sufficient.

The fry usually try to stay far away from the front end of their mother.

from each female would be kept separate. The F1, or next generation would be bred again brother to sister, keeping each line separate. You would continue this for approximately six generations. At this time you would cross a male from one line to a female from the other line, and vice versa. These again would be kept separate. If you watch what you are doing and pick the correct fish for breeders, it is possible to continue breeding this way for years and still produce quality fish. In reality this, too, is a form of inbreeding, but by running another related line to cross back into, you are able to retain many of the better qualities in fish that can be lost through a strict inbreeding program. With a little multiplication, you can see how this could get out of hand with regard to the number of tanks required to keep all these fish separate. The solution is to keep only a couple of droppings from each female. Occasionally, if I am using what I consider an exceptional male and female, I will keep more than two droppings. Recently I did just that and was lucky that I did. The young from the first two droppings were not much good; however, the fish from the third drop were exceptional. If I had followed my normal practice, I would have missed out on some excellent fish. So luck does occasionally play a part in breeding fish.

If the time does come when you wish to try and improve on your line of inbred or linebred fish, I would suggest using a male from another good inbred or linebred line. If you use a hybrid male, you could set

There can be a problem with long-finned males. If the gonopodium is too long, they can have a problem impregnating females.

This is a close up of a male guppy's gonopodium. There is a little hook at the end to grasp the female.

your breeding program back for a number of generations. With that statement we will go into the discussion of the hybrid guppy.

Outcrossing

Basically, a hybrid guppy is simply a fish or group of fish resulting from a cross of unrelated fish. The F1, or first generation possesses what may be called hybrid vigor, and may be outstanding fish in size as well as color. When trying to breed these fish one to another you more than likely will end up with junk. This is not true in all cases, but is usually what happens. Some breeders will use this method to produce their show fish, which works out fine for them while they have the two lines to make up for the hybrid cross. But it sure makes it tough on the person who buys these fish hoping to end up with the same quality fish as those they purchased. This is one of the reasons I would be reluctant to buy

Guppies left to breed indiscriminately will revert to wild type in a very few generations.

fish from a show unless you know the breeder and he will tell you if they were from hybrid or inbred lines. Fish from hybrid lines can be worked out with a good

Sometimes, several males will vie to mate with the same female. Male guppies are relentless in their pursuit of the females.

program of inbreeding, but it usually takes time.

BREEDING GOALS

At the same time, we have individuals who simply cross everything that they have in the hope of hitting the right combination and coming up with some good show fish. Once in a while it does happen and they win big for a year or so; shortly thereafter they seem to fade away as they try to reproduce the same cross that gave them the good fish.

I do not mean to imply that it is unwise to obtain another line in order to improve the finnage or color of your own fish. Often times this is the only way to get that extra little push your fish may need to boost them into the winner's circle.

When attempting to change the color of an existing line, it is necessary

A close-up shot of a female guppy giving birth.

to introduce a male having the desired color to a couple of your virgin females. He would then be bred to his daughters, and if at all possible to his granddaughters. The initial cross is unlikely to produce the color you desire, depending primarily on the compatibility of the females to the cross. By breeding him to the F1 and F2 females, some of the desired colored fish should crop up. Just follow your regular

The "gravid spot" is clearly seen on this young female.

A female guppy that is about to give birth.

breeding program from there on.

I have not consciously tried to breed large-tailed females in any of my lines. About once a year I will get a dropping or two that throw giant-tailed females. These tails are usually as large or larger than the males' and very colorful. People who see them can't believe that they have developed naturally, but such is the case. It must be the water. They breed

Guppy fry about one week old. Fed on baby brine shrimp, they will grow quickly.

Half black red female.

normally and the young are just as good as those of females with tails of normal size.

I guess part of the fascination in raising guppies is that you have almost complete control over their environment and development, which give you the opportunity to create something on your own.

Some time ago, in trying to develop an all-black guppy, it was found that an inheritable lethal gene was involved. As the black coloration was advanced toward the head area through selective breeding, the fish began to die off. Many breeders have tried this with the same results, so it would appear that we are more or less stuck with the $3/4$ or $7/8$ black-bodied fish. Even if one were able to develop an all-black fish it would only mean an extra point toward body color, hardly a critical factor.

When working with males with the characteristic snakeskin pattern, it is possible to use almost any virgin female and still carry the pattern into the next generation, which makes this one of the easiest fish to work with as the pattern is dominant in the males. A common weakness with this type fish is a short or small dorsal. I

Female guppies are usually considered bland and colorless. A good female will be quite attractive in her own right.

had developed a strain of these fish with a yellow and black snakeskin pattern over the entire body, caudal, and with a large, flowing dorsal. They won whenever I showed them, but economics forced me to discontinue this line. For years afterward I received calls from people who had seen these fish and wanted some of them for breeders.

At the present time, breeders of this type fish seem primarily concerned with size only, figuring perhaps that this will make up for the deficiency in other areas. What many people fail to realize is that when you increase the size of a fish, the pigmentation and pattern is spread over a larger area and tends to bleach out or dull with greater distribution. This is one of the reasons that the smaller German fish appear to be so brightly colored.

With some lines you can use color in the female's caudal as the determining factor as to which females to use for breeding. For example, in the red lines a female with a blue cast may throw the best fish. Keep track of which ones throw the best fish for future breedings.

A great number of caudal variations have been applied to the half-black fish, often with very striking results. My work has been primarily with the half-black reds. I have found it best to work with two different lines of the same strain. In most cases I will run three lines, two of the lighter bodied or gold half-black, and one line of the gray-gold body. Whenever I notice one of the lines losing a little color or size, I simply cross into one of the other lines, usually with good results. A

While pretty to look at, this is not a well-bred guppy by show standards. He is ideal for a community tank with other species of fish.

number of the breeders will use some of my fish to cross into theirs to improve the quality. The blue females in all four of my lines seem to work out quite well when crossing for a half-black blue. These are pretty fish and at times I think I would like to get into them but simply lack the tank space to do so.

If a neutral line is not available to you, females from a light green line seem to work out quite well for crossing. A good rule of thumb is to use fish of the same general color as those you wish to improve. It is sometimes possible through very selective breeding to eliminate many undesired characteristics. The major drawback is the amount of time invested and the number of tanks involved.

Initially, in the production of the type of guppy we have today, this work was necessary. It is much easier now to obtain another line from one of the established breeders to work into your line.

A tankful of males growing out until the lucky ones are chosen for breeding.

Half black red female with good caudal color.

DEFECTS

I have made it a practice not to use for breeding any fish with a discernible fault. It has been tempting, on occasion, to use a fish that was superior in most areas to his tankmates, except perhaps for a slight irregularity in his dorsal or even a mismatch in color. To do so would be to invite trouble further down the line and would require much reworking.

One unusual genetic problem I was unable to rectify involved a line of black-tailed deltas some years ago. These fish grew well, had excellent color in the dorsal and caudal fins, and were virtually disease free. Productivity and longevity were normal in all aspects, which usually would have made this an excellent line to work with, but there was one slight problem. If these fish were placed in a plastic bag and the outside of the bag got wet, in a matter of a few minutes all you would have left of the caudals were the rays; all else would have disintegrated. The males would be alive and still trying to swim with no tails. The first time this happened I thought that perhaps bad water was causing the problem, but subsequent testing proved otherwise. Different medications were employed

A tankful of colorful guppies destined for the wholesaler. These little beauties may not be show quality, or even well-bred, but they will bring great pleasure to the people who keep them.

BREEDING YOUR GUPPIES 129

Half black pastel veiltail.

AOC delta.

BREEDING YOUR GUPPIES

as a preventative with no results. Their diet was modified, again with no change. I didn't feel right about shipping or selling fish with this condition, so after a couple more generations with no improvement, I gave up on them. As this appeared to be a genetic weakness, possibly I could have resolved this problem through outcrossing, but again, I just didn't have the required tank space.

MAINTENANCE OF LINES

Usually, I make one cross among my lines once a year, and I am presently working with three red lines, three half-black red lines, four lines of blues, and two lines of purples. With only 66 tanks at this time, I really don't have

Half black blue delta.

Blue delta bred by Bill Orth.

any spare space to devote to additional crosses. In comparison to many of the eastern guppy breeders, I have two to three times the number of fish per gallon of water that they do. I have to do a lot of combining of different fish to a tank, which is not a problem as long as they are from differently colored lines. Also, different lines of females are kept together. The majority of my females are kept virgin to work back with if I need to. I will breed fish up to 12 to 14 months of age if I didn't have the opportunity to do so before then. The only thing I do is put them in a smaller container for breeding. Some breeders clip the tails of the large males in order to get them to breed. I did this just once.

It is not advisable to crowd fish as much as I do, as it cuts down on your

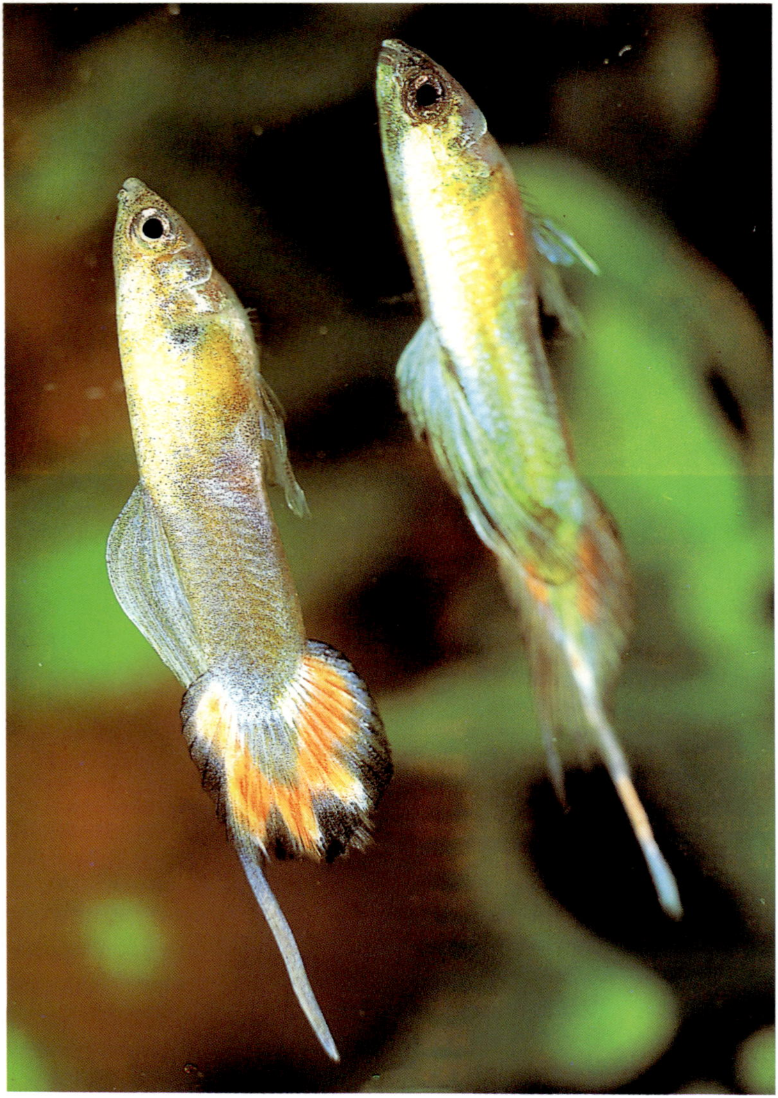

Facing page: *This type of fish is not recognized in the U.S. It could be called a "top sword pintail."* ***Above:*** *This multi-colored delta is no show fish, but that does not mean he is not lively and attractive.*

margin for error a great deal.

You will note that I haven't gone into the genetic structure all that much. I had more or less planned to make several charts, but as this has already been well presented in many books, including *Guppies: Fancy Strains and How To Produce Them*, TFH publication TS-122, it would have been redundant. Very basically, all I have been concerned with, as far as Mendel's findings apply to the guppy, are the percentages involved when crossing gold to gray guppies—or working with hybrids. It helps to know which colors are dominant and which are recessive, and also which body patterns are dominant. I would like to say that there are hard-and-fast rules, but such is not the case, due to the fact that there are very few, if any, "pure" lines. While we have standards to work toward for competition, many people are quite content just to raise what they think are nice fish. For the most part I have to go along with them—the fish are primarily there for your pleasure and enjoyment.

I breed primarily for the competition of the shows, but that is my enjoyment in consistently attempting to raise the perfect fish. I don't think I'll make it, but that's okay too. Each to his own.

TRYING NEW COLOR LINES

When I have a season where the breeding program is going along according to schedule and the lines are producing the quality of fish that I want, occasionally I will try a color cross just for something different. Two of my blue lines will drop some greens and purples along with the blue. So if I feel like running another color, it is possible to do so with this combination. By using a green male bred to one of his sisters with a light blue or greenish cast to her tail, I can come up with some very decent fish.

This way it is possible for me to add other colors to my set up, producing show quality fish, and still keep "in house" with my breeding procedure. As you work with your various

Red bicolor delta bred by Franklin Batell.

lines, you will find instances such as these where you are able to modify your basic colors to produce different ones. I have done the same thing with the half black lines crossed into one of the other basic color lines. With this type of cross you will often pick up some very interesting caudal color variations along with the larger flowing dorsal.

With a four-month turnover period it is possible to develop whatever color variations you desire in a reasonable length of time. This is provided you pay strict attention to picking the right breeders and the separation of the young males from the females. You do not want any prebred females if you are attempting to modify the color in a line. With this particular cross that I mentioned, the first dropping of the brother to sister cross will produce a good number of greens or purples, whichever way I decide to go. If one were working with unknown lines, it might be necessary to breed the father to a couple of his daughters to fix the color. All you have to do from then on is follow the normal line breeding practices.

BREEDING AGE

One failing that I share with a lot of other breeders is waiting too long before setting up breeders. In my case it is usually caused by not having enough tank space and I end up trying to breed fairly old fish, which on occasion hasn't worked out too well. Sometimes the older females will only drop a few babies that are not worth tying up a whole tank for. What I usually end up doing is putting them in with some other babies in another tank. The only problem there is that they were probably born at different times, which puts fish of a different age in the same tank. Then you usually have a growth problem. So if at all possible, try not to wait too long before setting up the breeders on the chance of getting few or no babies.

I have also found that if

This is the type of fish that is being commercially produced in Florida.

you are doing quite well with a particular line, there is a tendency to get careless and delay in setting up the breeders. When this happens, many times you end up using inferior fish for your breeders instead of the better quality fish you just watched slip away. In my own instance, if you will, I show against myself in that I constantly try to raise better fish than those I now have. Without the fish shows, I think it would be difficult for a person to maintain interest in just one line or color type fish.

SURPLUS FISH

Many times local pet shops will be glad to take your surplus fish off your hands. Usually, locally raised fish are in better condition than those shipped in from overseas. If weather conditions are bad, often shipped fish arrive in poor shape. You can sell them outright or trade them for the supplies you need. An honest dealer will treat you decently and give you a fair price for your fish. One thing is certain, you cannot raise fish for the wholesale price paid for the fish in Singapore; our costs are just too high here.

If you are going into guppies just to make money, I would suggest raising other types of fish. Hopefully, you can just about break even when you take into consideration the cost of equipment, food, brine shrimp eggs, and frozen shrimp. When you add in the cost of heating the fishroom, lighting, and water, it doesn't leave much room for profit. If you show our fish to any extent, either shipping to or attending shows, forget the profit factor, it just isn't there. However, when you also add in the enjoyable hours spent with your fish and the nice people you meet in the hobby, it does balance the books a bit.

CARE OF THE FRY

Perhaps after their parentage, the most important factor in raising quality guppies is the care they are given right after birth to the first three months of their lives. It is this period that usually makes or breaks a show-quality specimen.

It is generally a good idea to isolate the female a few days before she drops. Place

These are pintails. They are close to wild-type guppies. They are not shown in the U.S., but seem to be popular in the Orient.

her in a small tank, usually 5 gallons or less, or in a drum bowl. This serves two purposes: it prevents other fish from bothering her and also lessens the possibility of some of the babies being eaten. The females are fed well during their confinement with live baby brine shrimp, frozen adult brine shrimp, and light feedings of quality dry food. If using a drum bowl, the water is changed every other day with water taken from a healthy tank. It is advisable to place the bowls in areas where the fish will not be disturbed by your working around the tanks. You should not use fresh tap water for these bowls or for the confinement tanks; rather use conditioned tank water so she will feel comfortable and relaxed.

Usually a couple of days after the young are born, I transfer them to a 5- or 10-gallon tank with moderate filtration. As soon as they have made themselves at home, they are fed lightly with baby brine shrimp. It only takes a day or so for them to get the idea that when you approach the tank they are likely to be fed.

After a day or two of feeding them all the baby brine shrimp they can eat, I skip a feeding, and when they come to the front of the tank to eat, there is no food for them. An hour or so later, I add a light dusting of dry food to the tank and usually they will pick at it after a few minutes. This conditions them to eating dry food, and that's when the feeding program really begins.

A few of the breeders are bothered by cannibalism in their fish. I don't seem to be any more, possibly because of my feeding program, or by the fact that I destroy any female that eats her young. An exceptional female is given a second chance, but if she behaves this way again, she is eliminated. You do not have time to fool with fish that eat their young.

An average drop will usually be between 30 and 50 young, though I have had as few as 6 and as many as 130 from one female. Babies will vary in size at birth, but generally the very small ones will catch up in size after a week or so.

The basic reasoning behind the small tank for the young fish is in order to

A Japanese pintail.

have a large concentration of live baby brine shrimp in a confined area. This makes it much easier for the babies to feed on a continual basis. Using a moderate flow of air to the filters provides the necessary oxygen for a variety of functions without boiling the water with air and disturbing the fish.

The use of inside corner filters, with the tops removed and filter floss as a filter medium, traps the uneaten live shrimp and provides snacks for the baby fish throughout the day. Ideally, the stomachs of the young, growing fish should be distended at all times, full of good food. While it is possible to grow good fish without the use of baby brine shrimp, much better results are obtained through the use of this "ideal" baby food.

It is important to have the baby tanks placed in an

area where you are able to keep them under constant observation, with good lighting, generally on the top tier of tanks. This also places them in the highest room temperature zone, which tends to speed up their metabolism, which causes them to eat more, and the more they eat the faster they grow.

While it is possible to determine the sex of young fish when they are 3 to 4 days old through the use of a strong light, there is no advantage in doing so. In fact, to my way of thinking, there is a distinct disadvantage in separating the males from the females until it is absolutely necessary. By keeping them all together, the young females, by their increased metabolism, eat more and so induce the males to eat more also.

I have found it best to begin separating the fish at about three to four weeks of age. As soon as the young male's gonopodium starts to develop, I net out the females and place them in a different tank. The reasoning behind this is, if you try removing the males, there is a possibility you might miss one or two, and these could mess up your virgin stock. By taking out the females, missing one or two will not make any difference. The females can easily be identified later by their gravid spot, a dark area behind the anal fin.

It should be noted that as the fish get older, they are transferred to a larger tank. Usually by the time they are three weeks old they are placed in a 15-gallon tank. At about one month of age, all of the females will have been removed. The young males are left together from six weeks to three months

This fish is called a red snakeskin in Singapore.

of age. If these fish are to be raised for show purposes, the population is cut down to approximately 1 to 1½ males per gallon of water. This is not a hard-and-fast rule, it being necessary to experiment to determine the best growth rate per tank size under your own conditions. Also, at this time you may wish to cull any young males or females that do not look quite right to you. This could be in the form of poor body shape or even deformed bodies, or improper fin development, such as elongated rays or ragged edges on the caudal. In my own tanks I do very little culling, which makes it possible to raise quite a few different strains with

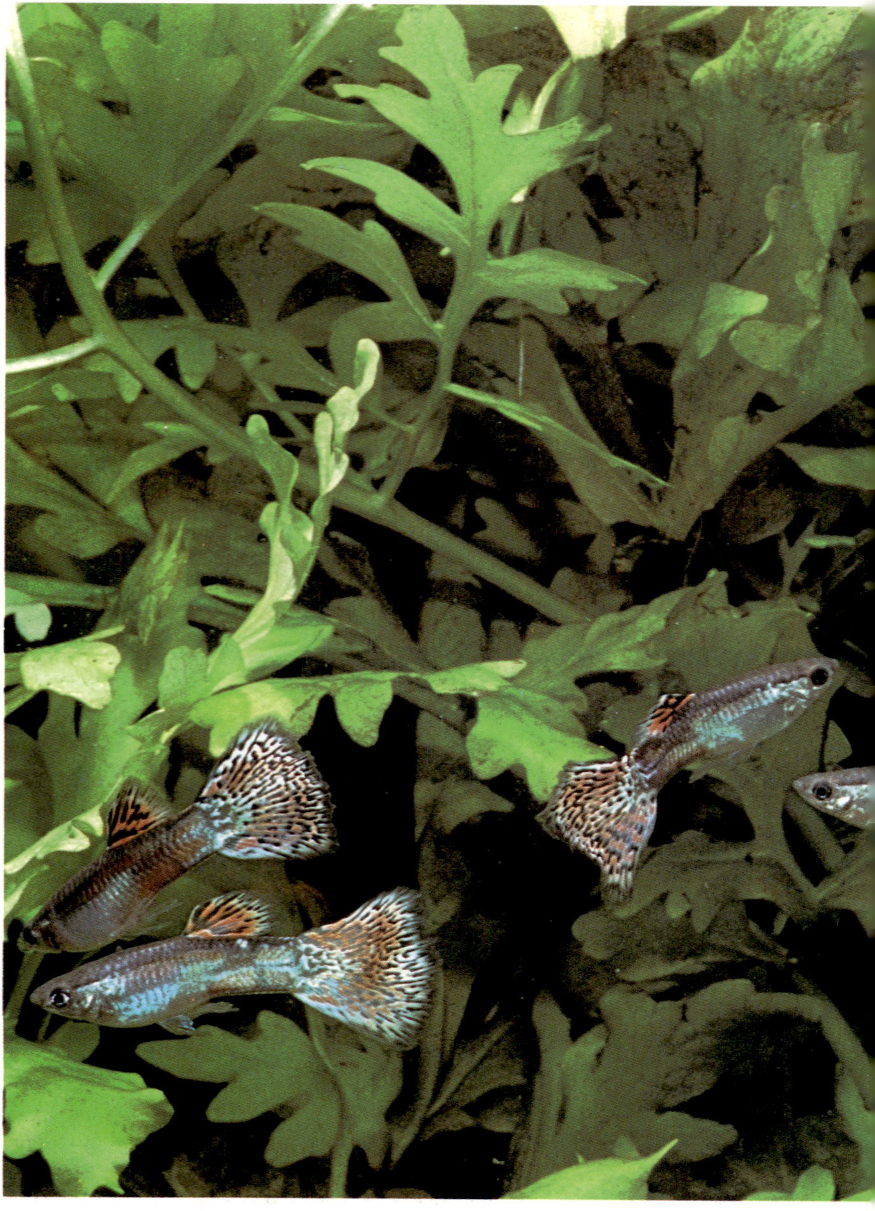

This charming planted tank will support a self-sustaining guppy population for many years.

relatively few tanks. With the high percentage of good fish to work with, it is not necessary to keep large numbers of fish to get a few good ones (the reason for culling).

Usually the late-developing males turn out to be the best fish, so don't be in too much of a hurry in picking what you think will be your next breeders.

Water conditions seem to be a determining factor in the development and coloring-up of the young males. Fish, which in your tank may be fully colored and sexually active at five weeks of age, may not even start to color at two months in another's tank. The longer they wait to become sexually active, the bigger they usually grow.

Temperature is a prime factor in the development of any fish. As fish are cold-blooded, their metabolic rate is determined by the water temperature. By raising the temperature, you speed up the metabolism, causing them to be more active and to eat more. It would seem logical to raise the temperature way up, throw in lots of food, change the water often, and violá! Big super guppies. Well, not quite. While it is true that the fish will mature faster at higher temperatures, it demands a price. The main problems seem to be shorter life spans and increased bacteria, both good and bad. The latter requires close observation of the tank conditions, especially for the males. If the bacterial balance goes too far out of kilter before you catch it, the males' caudals can deteriorate in a couple of days, undoing all your work.

Facing page: *Young multi deltas in a planted grow-out tank.*

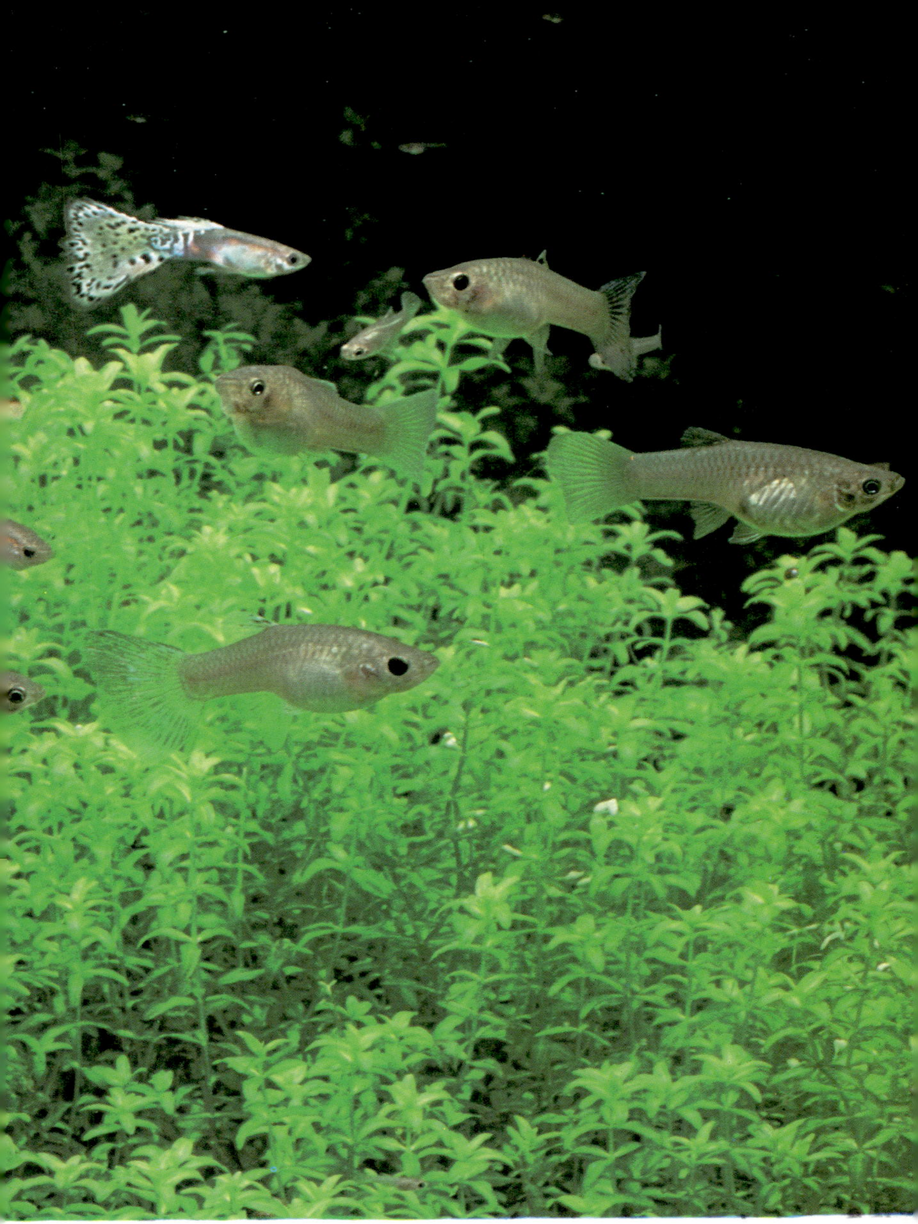

Some years ago, when I had individual heaters in each tank, I ran a series of tests on temperature variables. The lower tanks were running around 72°F. and the top ones were 84°F. Feeding and water changes were approximately the same, although the fish in the top tanks seemed to eat more; all were given as much as they would eat. At about one month, the fish in the higher tanks were starting to pull ahead in size and continued to do so. At four and one-half months of age I was able to show these fish and win with them. The fish in the 72°F. water were nowhere near ready for show. I continued showing these fish raised in the higher temperatures, on and off, until they were

Half black pastel delta bred by Jim Alderson. The caudal and dorsal are well shaped. The body and dorsal color is good but the caudal color is only fair because of the spots and the yellowish area by the peduncle.

This is what was affectionately called a Flamingo. It is actually a gold red-tailed delta. It could use a bit more color at the back end of the dorsal.

seven and one-half months old, when they started to deteriorate. From then on, it was downhill until they died at about 12 months of age.

The fish that had maintained between 74°F and 76°F started getting prime between seven and eight months of age, their bodies developing first, then their caudals and finally the dorsals. These remained in showable condition up to 14 months of age. The fish in the lower tier were at show size around nine months and were still showable at 18 months; these fish lived about two years. For some reason, the gold guppies tend to live up to three years, and I have shown some at two years of age. I guess the genetic makeup of this type of guppy contributes to the longevity.

One thing should be noted: although the fish kept at the higher temperature grew larger faster, the ones in the cooler water eventually caught up with them, and in some instances surpassed them in body size. So you have a choice of fast growth, moderate growth, or slow growth, whatever you require.

I have tried to develop my strains into a slow, continual-growth type of fish. I receive calls from other breeders to whom I supply fish, saying how surprised they are that the fish I sent them are still growing at almost a year of age.

It should be noted, however, that the size of the fish will to a great extent be determined by the genetic makeup of the strain. Regardless of the ideal conditions and perfect food you provide, the fish will only grow to their maximum genetic potential. If the strain is small in nature it becomes almost

Half black AOC delta male.

BREEDING YOUR GUPPIES 151

to increase the growth factor. However, it would be advisable to make the cross with an established inbred or linebred strain. This will increase the chances of obtaining the required size development without sacrificing too much in the way of quality, which could happen if the cross were made to a hybrid line. It is possible to place different types of males together if your dropping

Above: This is an excellent snakeskin other than a little loss of color at the top rear of the caudal and a slight shade of difference in the dorsal.

Below: Snakeskin with a delicate pattern.

is not large enough to warrant tying up a tank for just a few fish. For instance, you can place reds or half-blacks in the same tank and easily tell which line they came from. You may also do the same thing with females.

The only problem I run into is with my blues, as I run four separate lines. It is necessary to keep them all separate. One other thing you should not do is mix different age or size groups of fish. The reason for this is that there seems to be a growth inhibiting substance given off by fish that will definitely affect the smaller fish if they are in the same tank with the larger ones.

Frequent water changes will somewhat alleviate this factor, but reduced growth will be readily apparent if you do try to mix the different sizes. I am still not completely convinced of this growth inhibiting substance theory. As a case

Longfin multi which took a first prize in an Austrian show.

in point, if any young fish are left with the parents in the breeding tank, almost without exception they will be larger than their brothers or sisters after a week or so. This could just be the availability of more food with less competition for it. A reverse situation would be in a tank with a large population, the bigger fish would get more than their share of

Above: *Multi delta. The color in the caudal is a bit weak and the dorsal doesn't match.*

Below: *Multi delta with good caudal but only a fair match between dorsal and caudal.*

the available food and the smaller ones would suffer.

If enough tanks and time were available, it might be interesting to check into this a little bit further. I never seem to have enough of either.

One thing I have noted because of the arrangement of my tanks is that when I am required to reach across one tank in order to feed another, it is advisable to feed the tank you are reaching across first. For some reason with some lines of fish they get upset very easily, but if fed first, you can reach across them without disturbing their feeding pattern.

Multi delta male bred by Jack Longer. Nice fish with good finnage and color.

Young blue with smooth delta tail with excellent body shape and color. The body colors do not show well at shows most of the time because of the lighting.

Sometimes you will get a young male that seems to grow unusually large. At about three to four months of age, when the caudal is normally starting to really fill out, this fish's tail doesn't seem to grow any more. Even though he has the male sex organs, at least externally, and usually some degree of body color, which is as far as he or it will go. This fish is what we term a "mule" and will not breed. The other males in the tank seem to realize this and will chase him as they would a female. The fish itself displays somewhat passive, feminine behavior. You might just as well eliminate any such fish as it will not develop into anything.

Because of the additional

This fish has superb body color, perhaps the nicest I've ever seen. It's too bad the caudal doesn't match.

amount of food you are trying to pump into the baby fish, very often you will find mulm, or uneaten food in clumps on the bottom of the tank. It is advisable to net or siphon this out before it gets to be a problem. A little bit seems normal unless it gets out of hand. If the tank remains relatively clean, it means the filter is working properly and you are feeding the right amount of food.

BREEDING YOUR GUPPIES 157

Above: *Young half black AOC male. Pastel caudal with polka dots make for a very attractive fish. He could use a few more dots in the dorsal, but still a prospective breeder.* **Below:** *Half black AOC delta.*

A partial view of Stan Shubel's collection of guppy trophies!

Showing Your Guppies

Assuming you have followed (or not followed, as the case may be) all of the methods outlined in this book, you've produced some pretty good fish. In many of the metropolitan areas of the country there are aquarium clubs that have shows. You may wish to exhibit your fish at one of these shows. The only problem is that at this type of show the guppy classes are quite small and offer little competition. Thus, they do not give you much of an idea of how good your fish really are. If at all possible, I would suggest attending an official International Fancy Guppy Association (I.F.G.A.) show, even if you have to travel to do so.

This will give you the opportunity to see some good fish as well as meet and talk to some of the guppy fanciers who make up these shows. I have had dealings with many other types of aquarium clubs as a guest, a member, a speaker, or a judge, and have never met a nicer bunch of people. If you have questions or problems, almost any of them will be glad to help you out.

Also, when it comes time to judge, you may sign up as an observer and follow a judging team around and see what the judges are looking for in picking the top fish.

When I first started judging in my own region, usually two other major judges were called in. Normally we would split up the classes and point each fish individually. Depending on the size of the show, this could take up to twelve hours. By the time you got to the last of the fish, you were not as sharp as you should be. Many good fish and their breeders suffered because of this. Very few of the shows were judged the same way and frankly, some of the judges were rather poor.

Through the efforts of a lot of dedicated individuals and hours and hours of work, we came up with a set of standards for the judging of the fish, running of the shows, and for the judges themselves.

In order to be a qualified I.F.G.A. judge, you have to observe a required number of shows, and assist judging at a required number of shows after you have passed a written test. Then you will take a visual test, placing a group of males in the proper order using the

Matched pair of red deltas, excellent fish that took best of show. Notice the nice clean lines of the caudals.

Half black AOC delta. Another first place fish with good dorsal and caudal match. Very good body shape and color.

official point system. Once you have passed all the requirements, your name is brought before the judging board for approval. Only then are you issued a judging card. Seminars are held at least once a year, and you are required to attend one at least every other year to retain your accreditation. The visual test is very hard to pass; many individuals who know the point system forward and backward cannot pass the visual test, and so will be unable to receive their qualified judging card and will remain assistant judges.

This is a shame, as some

of them are otherwise excellent judges. Everyone does not necessarily agree with the regulations and rules in their entirety, but the judges follow them nevertheless. In judging it is hard not to have a personal preference as to color, type of caudal, or even shape of dorsal, etc. When judging at a show, personal likes should be set aside and judging done only by the standards. At any I.F.G.A. sanctioned show, you will be virtually assured that your fish will be judged correctly and fairly. *Tropical Fish Hobbyist* magazine lists many of the shows held throughout the year as well as contact parties for more information on the I.F.G.A.

When at all possible I like to attend the shows in person. It has gotten to the point where even if I sweep a class or take best of show,

TYPICAL FISH SHIPMENT BOX

Half black blue delta.

it doesn't mean all that much unless I can see how good the competition was. This is the primary reason I don't ship fish to shows all that much. However, if you are going out for a class championship and are unable to attend each show, you must ship.

SHIPPING SHOW FISH

If you plan on sending a large number of entries or if cost is no factor, the best way to ship is air freight. I send express mail with next or second day delivery.

Preparing the fish for shipment is very important if you expect them to do well and return in decent shape. Do not feed them for a minimum of 24 hours before shipping; 36 hours is better. Take one to one and a half cups of water for each fish from their own tank. Do not add any fresh water. Use two plastic bags, one inside the other. Some breeders invert the bags, but I have not found it necessary to do so.

After adding the water and the fish, spin the bag holding the top so as to capture an air pocket. Twist the top of the bag and bend it back over, and fasten it with at least four rubber bands. The reason for the air pocket is to get a wiping action for the transference of oxygen molecules.

Make sure you use a good grade of bag, one that does not leak. Otherwise, the chances of your fish making it to the show in good shape are very slim. As long as the outside of the bag is dry the fish will usually be okay. I have had them stay in the bags for 11

A young half black AOC female. Notice the good color in the finnage even at this age. She also has a very good spread to the caudal.

SHOWING YOUR GUPPIES

A tank of my three-month-old gray bodied half black pastel AOC males. Because of the varied pattern in the caudal, it is difficult to pick a matched tank entry.

days with no ill effects when shipping out of the country. It is advisable to put only one fish per bag, except when using air freight when many can be put in a single large bag. There will be some expansion of the bags during flight, but so far I haven't had any trouble with the way I pack the fish for shipping.

After bagging the fish are packed into a styro container which is placed in a heavy cardboard box completely sealed with tape. Boxes are marked "Live Tropical Fish, Please Keep Warm." When shipping to foreign countries, use their language as well as English.

When your fish arrive at the show site they are placed in conditioned water, and usually given the best care possible. After the show is over, and hopefully you will have won a couple of trophies, your fish will be sent back to you.

Upon the return of your fish, place them in a tank of conditioned water. To the water I usually add 250 mg. of Tetracycline per five gallons. I add 2 to 3 drops of formalin per gallon of water as well.

The fish have been through a great deal of stress and strain in the past few days and are generally in a weakened state. The medication is added to prevent any disease problems. After observing the fish for several hours and provided they come to the front of the tank looking for something to eat, Feed them very lightly. The next day, if everything appears normal, you can return to your regular feeding program.

When a couple of more days have passed, and all seems well, you can place them back into their original tank. Better safe than sorry. Even with the best of care at both ends, you can expect an approximate loss of 20% of your shipped entries, so you should plan on this if shipping to many shows. Unless they develop a disease, the fish recover very quickly and resume their normal habits.

When attending the show in person, take all conditioned water for the entries. I have tried different combinations of fresh and aged water, different medications, and have found that the best way is to take water from fairly clear, healthy tanks with no additives. Any time you add fresh water, there is some shock involved to the fish, and the show is no place for shocked fish.

The fish are carried to the show in two-quart drum bowls covered with a single thickness of plastic bag

This young half black yellow delta was bred by Paul Evans. As a rule, most of the fish of this type do not get quite as large as some of the other show fish. Nevertheless, they do very well because of their excellent color match.

cover, well-sealed with large rubber bands so that the water doesn't slop all over.

I have placed as many as eight fish per bowl, but if travelling a long distance, I try to keep it down to four per bowl. On arrival, the fish are placed on the bench. Some are transferred to a display tank; all are left alone for about 15 minutes until they color back up. If there are any questions in my mind as far as color goes, I take them over to the show bench with its black background and check the color.

SHOW CLASSES

If you are a newcomer to the show, the show chairperson will have someone help you in the classification of your fish, advising you as to the color and caudal shape of your entry. The fish do change color at the shows, so what you had at home might have been a nice green, only to end up disqualified as a blue under show conditions. Usually under the show rules they will list the type of lighting to be used and you can be guided accordingly.

In the male classes, you may add a female if you choose; she will not be judged. Some exhibitors feel that the female will get the male to display better. I do not like to do this because after a day or so with no food, that nice flowing tail on the male will often tempt the female to take a nip out of it. Quite often that will be enough to knock your entry out of competition.

I do not color feed my fish, as the food I use seems to give the fish enough natural color. Very often, people who show blacks will color up their fish artificially, but they are getting to the point where it is no longer necessary to do so. Some foods, and I guess mine would have to be included, contain ingredients and possibly certain vitamins or minerals that enhance coloration to some extent.

A few breeders use different types of natural foods in an effort to increase the intensity of color in their fish. Carotene and paprika are sometimes used in the attempt to bring out the red

Red deltas from my line showing good caudal and dorsal match. They should be ready to show in about one month.

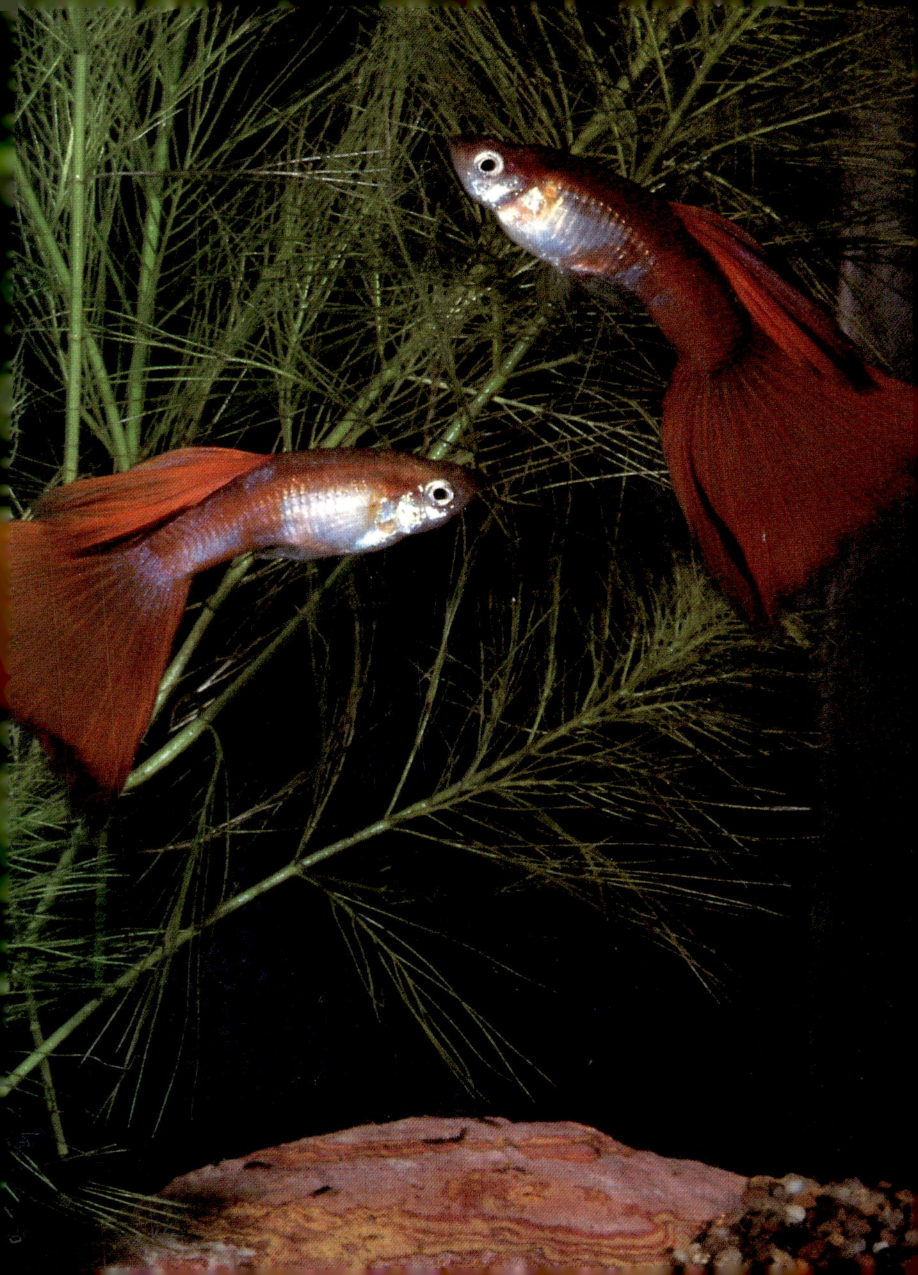

color. At shows, red guppy breeders are surprised that my fish are so good a red color without using something to bring it out. I found a number of years back that most of the color enhancements were only temporary, and the way to go was to improve the fish through selective breeding. When I enter in a color class usually I will have fish from two or three different lines. As there is a slight variation in the shades of color, one may show up better than another depending on the lighting. At one of the shows in blue class the fish placed 1st, 2nd, 3rd, and 4th, each one from a different line. Also, some judges seem to favor, say, a medium blue over a dark blue or vice-versa, so if you have fish displaying the different shades, you increase your chances of winning.

Often, fish that were very dark in your tanks will

Young half black red gold male with good matching dorsal and caudal. Body color will get darker as the fish ages. He is about 3 1/2 months old now, a possible future breeder.

One of my dark purple deltas, showing very good shape as well as color match. A few years ago, some exhibitors would color feed this type of fish and put it in a black class, usually with very good results.

brighten up when placed in the show container. As a rule, I do not take fish that are light in color unless they are very bright or intense.

My fish are very active at the shows, which is a definite plus when it comes to judging. A lot of the fish are inactive or frightened and stay at the bottom of the container, which makes it difficult.
They are not allowed to place anything in the show container to move the fish, so they can only judge what they can see. A reasonable effort is made to get them up and swimming, but if they fail to do so, you can forget about them placing.

Often when I am attempting to modify a particular color I will take some of these fish to a show for a color check against the other entries. If they win that's fine, but I really want to see how they stack up against the fish being shown in that class. You get a somewhat narrow view of your fish unless you can see what they look like against others not of your line.

In entering the matched male classes, an important factor for some reason is that the bodies are all the same size. Sometimes I feel that too much emphasis is placed on this. Basically what you do is judge one fish and point him in all three areas and compare the others to him. For instance, the color match on the caudals and dorsals should be the same; body color, body color patches or dots should be the same. If there are any color variations in the caudal, in the form of different shades or patterns, it should be the same in all the fish. Size of caudals along with shape should be equal. The same should hold true for the dorsals.

Blue delta line. An 8-month-old show winner. This fish is close to a super delta. Note the length of the dorsal and the match.

Half-black blue delta male.

Usually if I plan to enter the breeder class, which consists of five matched males, I will bring at least seven matched males, and after they have had time to color up I will pick the best five of the group. Normally in the two matched male class you are able to color match them in your tanks at home and they will hold fairly true in color once you get them to a show.

The classes we have the most color problems with are the green, blue, and purple classes. Lighting is such an important factor in bringing out the proper color that these three classes are generally placed in a section of the show area where you have some frontal light. Front sunlight would be ideal, but at most of the shows it is impossible to have. We tried handheld lights of various design, but depending on the angle they were held, as well as the angle the fish were viewed from, you could get three different colors from one fish. Most of the shows will use overhead fluorescent lighting placed above and between the show racks—not perfect, but they do a fair job.

Even though my fish have won hundreds of awards, I think my biggest thrill came at a major New York show where my blues took all twelve places in the class along with best of show single, best of show matched males, and 1st place breeder male. I don't think this had ever been done before or since.

Usually I try to enter my fish a couple of hours

Solid snakeskin delta bred by Captain Frank Orteca. Well-shaped body and caudal with a good snakeskin pattern on the body. The dorsal is a little small and off-colored.

Young blue breeder displaying a nice delta tail with good body shape. The dorsals on the blues seem to develop slower than the dorsals on the red lines.

before the judging is scheduled to start. This gives them a little time to settle down and color up to normal. During the actual judging, there is quite a bit of moving the jars around, and if the fish has just been entered in the class, his color may not be what it should be. Your water should be as clear as possible in the show container; if colored too much, the entry could be disqualified. Also, the judges like to look through nice, clean water. Minor points, perhaps, but it could mean an extra point or so for your fish. As is so often the case, one point can make the difference between placing and not.

As in any competition, it is to your benefit if you know what the judges are looking for. If you are unable to attend a show, you can request a qualified judge to check your fish after the competition is over and note areas your entry could have been improved.

At every show I have been to, people ask me to check their fish and offer suggestions for improvement. Apparently this service by myself and other judges has been useful, as many of these people are now in the winner's circle.

There are a number of individual breeders who, while not making a big splash each year at the shows, continue to raise good fish. They may only

Green delta bred by Darell Johnson. Smooth even color in caudal; dorsal is a little small and slightly off color.

Green delta bred by Jim Alderson. Good dorsal and caudal color match. This is a nice large fish with good dorsal size. The pattern in the dorsal is not desirable because of the solid color caudal.

show a few fish a year, usually doing quite well when they do. We also have breeders with a lot of tanks and a lot of fish who go after the class, and male or female championships. In order for the hobby to survive and advance we need both of these types.

JUDGING

For various reasons, many of you will not be able to attend an I.F.G.A. show, so in order to give you some idea of what a judge will look for in a show fish, we will cover the point system, and using these points, will simulate the actual placing of a male and female guppy.

Regardless of which point system you employ, I.F.G.A. or the European standards, allocation of points may vary somewhat, but the results will be very close. This method will not be 100 percent accurate for obvious reasons, but will give you a good guideline and insight as to how a judge views your fish.

The fish below are judged using the present I.F.G.A. point system. The goal is 100 points. Beside each category is the total number of points possible in that area, i.e. body size, 8 points. For complete information, contact the I.F.G.A. and request a judging blue book.

First, we will judge the male.

BODY SIZE
(Value: 8 points)

You would start with the body size. This is the reference point from which we start. For whatever

Good AOC fish bred by Jim Alderson. The color match between dorsal and caudal is good, almost a pastel white.

SHOWING YOUR GUPPIES

I.F.G.A. POINT SYSTEM

	MALE			FEMALE		
	Body	Dorsal	Caudal	Body	Dorsal	Caudal
Size	8	8	11	11	7	11
Color	8	8	11	4	6	11
Shape	5	5	10	11	4	10
Condition	4	4	5	4	3	5

DEPORTMENT 5 SYMMETRY 8 TOTAL 100 POINTS

length the body is, the caudal fin should match in a 1:1 ratio. As I have the advantage of looking at the fish, I will assign a size of 6 out of a possible 8. Once you have this figure, everything else should fall into place.

BODY COLOR
(Value: 8 points)

Next would be the body color; ideally it should match the caudal and dorsal fins. As you can see, it is very close, with the color extending all the way up to the head. The chest area does not have the red color, so this would be considered an absence of color. No points are given for this area. This would also hold true for any off-color, such as blue or black spots in a red fish. Consequently, as approximately ¾ of the body is the desired color with good intensity and density, you would give this fish a 6 for body color.

BODY SHAPE
(Value: 5 points)

Going next to the shape of the body, you have 5 points to work with. Excellent would get 5 points; good would get 4 points; average would get 3; fair would get 2; and poor would get 1 point. This particular fish is fairly well-rounded with a good peduncle area, not chesty. It has a very slight hump behind the head. For this reason, it would only be classed as fair and receive 3 points for shape.

BODY CONDITION
(Value: 4 points)

Condition-wise, the eyes look normal; no scales are missing; gill plates look normal, so it would receive 3 points for condition.

DORSAL SIZE
(Value: 8 points)

Next, we will go to the dorsal. Ideally, for a delta, it should be one unit high by three units in length. This one is close. For size, this dorsal goes well beyond the peduncle area and is approximately the proper ratio, so you would give it a 6 out of a possible 8.

DORSAL COLOR
(Value: 8 points)

Dorsal color is almost a perfect match to the caudal with a nice even distribution. I would go at least a 6, possibly a 7.

DORSAL SHAPE
(Value: 5 points)

The dorsal shape looks slightly folded over at the top, which was not the case. I would give it 3 points for shape.

Judging fish, male. Red delta line with very good medium red color with excellent match. The body should be a little larger. Note how close together and fine the rays are in the caudal.

DORSAL CONDITION
(Value: 4 points)

Condition looks good, with no holes or splits or extended rays. It would receive 3 points.

CAUDAL SIZE
(Value: 11 points)

Now we will go to the caudal. It should match the body in length. Size is determined by the total area in proper proportions. The caudal could be slightly larger, but is close to the 1:1 ratio. I would give it an 8 out of a possible 11. You will note that as there is a greater total area in the caudal, more points are allocated.

CAUDAL COLOR
(Value: 11 points)

The caudal color in this fish is excellent with very little variation. Density and intensity of color is about as good as I've ever seen in a red fish. For this reason, I would go 10 points out of a possible 11. Always keep one point for improvement.

CAUDAL SHAPE
(Value: 10 points)

On the bottom of the caudal, about one-third of the way back, the line is broken and not as even as the top edge, a small chunk is missing on the trailing edge near the bottom and the angle from the peduncle is not ideal, so the most you would give it for shape is 5 out of the 10 points.

CAUDAL CONDITION
(Value: 5 points)

Condition of the caudal is not too bad, other than the one chunk missing. The natural scalloped edge would not be penalized. I would go 3 points out of 5.

DEPORTMENT

As to deportment, the fish is displaying itself quite well and appears to have no trouble swimming, so it would get 3 points.

SYMMETRY

Symmetry is the overall pleasing appearance of the fish in the proper proportions. Certainly the color match is pleasing; body shape is fair; caudal shape, fair; dorsal shape, fair; overall 1:1 ratio is good. Taking into consideration the weak points along with strong points, I would go at least 5 and possibly a 6.

As you can see by judging rather critically,

SHOWING YOUR GUPPIES

this fish would receive 70 to 72 points out of a possible 100. Another judge may point this fish in the 80s. As long as all the other fish are judged equally and with consistency, it makes no difference if you judge high or low. All things being equal, at a regular show this fish would likely win his class and place somewhere in the best of show standings.

When judging female guppies, the points are allocated somewhat differently. The ratio of body to tail is 2:1. In other words, the body should be two units in length to one unit for the caudal. Also, more points are given for body shape; less for the dorsal. Caudal points are the same as for the male.

BODY SIZE
(Value: 11 points)

On this half-black red female, we will again start off with the body size points. Out of a possible 11 points, I would give her an 8.

BODY COLOR
(Value: 4 points)

As there are only half the body color points as compared to the male, this female would get 2 points. She shows the half-black body color as required, but it is not really great.

BODY SHAPE
(Value: 11 points)

The shape of the body is pretty good; top is nicely rounded; stomach area could be slightly fuller. The peduncle area looks a little thin, but this could be caused by the dorsal making it look thinner than it is. I would give her an 8 out of 11.

Judging fish, female.

**BODY CONDITION
(Value: 4 points)**

Her condition looks good with no visible faults; give her a 3.

**DORSAL SIZE
(Value: 7 points)**

The dorsal is a good size for a female and would get 5 points out of 7.

**DORSAL COLOR
(Value: 6 points)**

The color of the dorsal is off when compared to the caudal, with only a trace of red, so the most she would get is 2 points.

**DORSAL SHAPE
(Value: 4 points)**

The shape of the dorsal is good for a female and would receive 3 points.

**DORSAL CONDITION
(Value: 3 points)**

Condition is good also and would get 2 points.

CAUDAL SIZE
(Value: 11 points)

Going to the caudal next, we determine it is half-black red, as there is at least 51 percent red color in the caudal. The size of the caudal would be an 8, possibly a 9.

CAUDAL COLOR
(Value: 11 points)

The color is not usually as dense or as intense as the males' and will be judged accordingly. Color distribution is good throughout the caudal with a blend rather than a solid bright red as was true in the male we judged. I would give this female 7 points out of 11.

CAUDAL SHAPE
(Value: 10 points)

The shape of the caudal is very good for a female with no real faults, so I would go at least 8 points.

CAUDAL CONDITION
(Value: 5 points)

The condition of the caudal is good with no splits or damaged positions, so I would give it 4 points.

DEPORTMENT
(Value: 5 points)

She displays herself well, holding the fins erect with good movement. I would give her at least 3 points.

SYMMETRY
(Value: 8 points)

This is a rather pleasing fish to look at. The body could be slightly larger overall. The dorsal is of good size, but could be better in the color match. The caudal is nice. I would go a 5 out of a possible 8. This female would do quite well in her class.

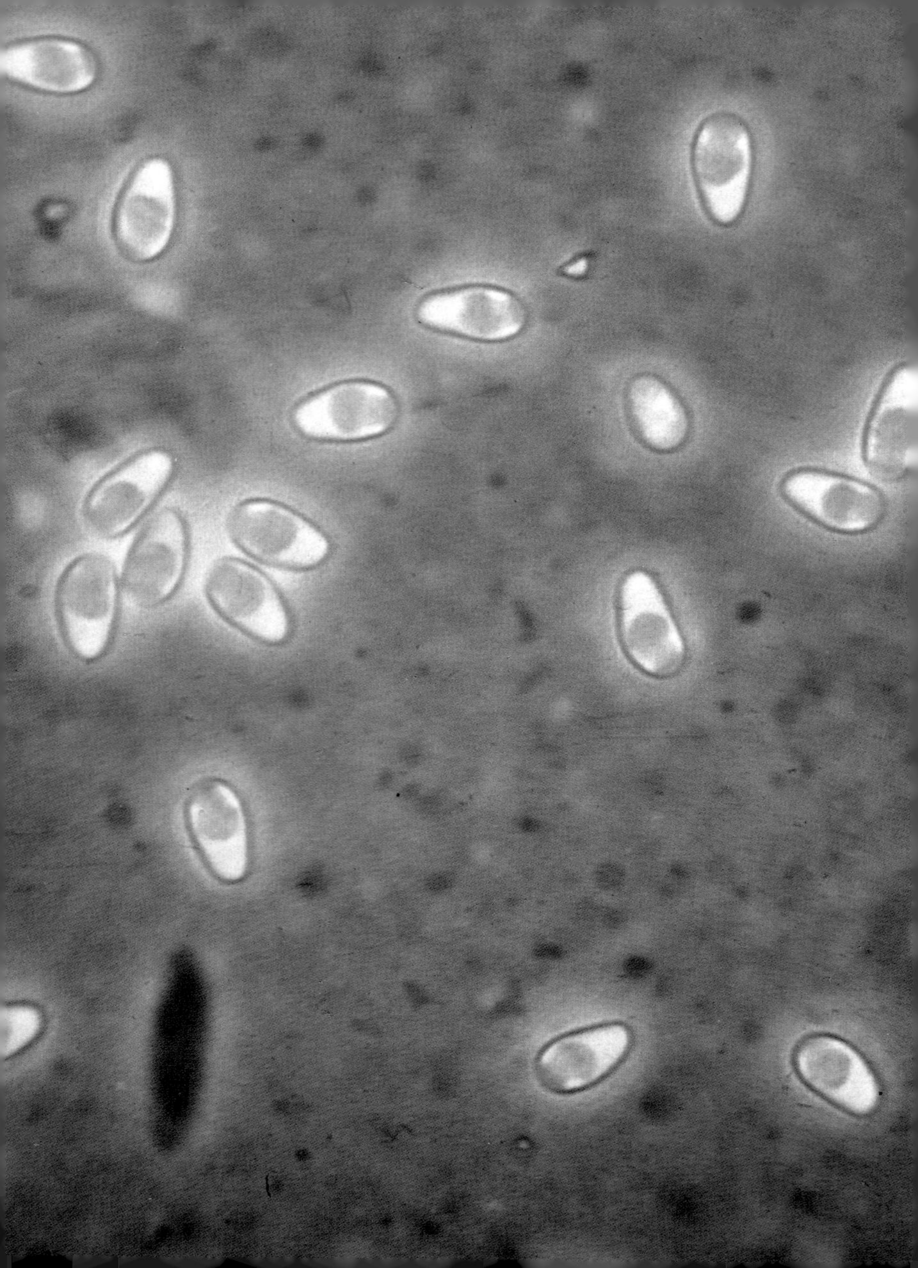

Guppy Health

The old adage "An ounce of prevention is worth a pound of cure," is perhaps never more true than in fishkeeping. Whereas you have very little control over your own environment, that which you exercise over your fishroom is almost total.

Consider the following: you provide all the food, all the water, lighting, level of temperature, and to a great extent, the amount of oxygen available in the water. So if your fish develop problems, you usually don't have to look too far for the guilty person.

Normally, fish kept in good tank conditions, with clean filters, and fed a well-balanced diet, contract very few diseases. Overfeed a couple of times, get lax on water and filter changing and suddenly you have problems.

With both sexes, one of the first symptoms is a general listlessness and lack of interest in food. If feeding dry food, they will usually swim under the food without attempting to take it, and will only pick lightly at the live brine shrimp. At this point the fish and the tank should be closely examined. If nothing

Facing page: Spores of Plistophora hyphessobryconis. *To make an absolutely correct diagnosis of the diseases that affect your guppies, you would have to have a microscope and very good references!*

seems to be wrong with the tank, filter, and no visual problem with the fish can be seen, the best thing to do is to stop feeding this tank for several days. If there is no major problem, the fish will snap out of it and be okay. On the other hand, if you fail to observe this condition and continue to drop food into the tank, you stand a good chance of losing a large portion or all of the fish in the tank. After several days the fish will start to get thin in the

Malnutrition induced lordosis, or curvature of the spine.

GUPPY HEALTH 189

Male guppy undergoing a sex change. The belly is filling with eggs.

body and from this point on they gradually deteriorate until they die.

The big-tailed fish we are trying to develop are a far cry from the wild natural guppy. The wild guppy, with it's small caudal, does not have the problem with the blood supply reaching the outer edges of the caudal as our show fish do. An inadequate blood supply will cause a weakening and gradual breakdown of tissue, which invites bacterial infection. There

A very advanced case of bacterial tail rot.

Tail rot.

is a natural slime coating on the fish which protects it from many diseases. If, through carelessness, we let the tank go bad, which in turn causes a bacterial buildup, a chain-like reaction occurs as the fish's defense mechanism breaks down, leaving the fish in a weakened condition and susceptible to any prevalent disease.

There are ways to lessen the likelihood of tail rot, one of which is to cut down on the amount and frequency of feedings to hinder the possible

GUPPY HEALTH

imbalance of the bacterial level in the tank. Lowering the temperature also retards the development of some bacteria. Keeping the tank clean of the slime that develops on the sides and the bottom of the tank will also help.

The use of salt will help promote the maintenance of the protective body slime but is of little benefit in preventing fin and tail rot. One drop per gallon of water of 37% formaldehyde (formalin) will also keep the bacteria

The raised spots of ich.

Saprolegnia, or fungus. Fungal spores are always present in the water, but are not able to attack a fish unless it is damaged or otherwise unhealthy.

at a reasonable level. (It is noted that the idea is not to destroy all the bacteria in the tank, only to keep it at acceptable levels.) Good bacteria will not harm the fish provided they are in healthy condition. If you see the water starting to cloud up, stop the feeding until it rectifies itself. I might add also that older, larger fish do not require as much food as do younger, growing fish.

You might wonder how all of these diseases originated in your more or less closed environment. Many are already present in your fish tank, but do not manifest themselves as long as the fish remain in good condition. Others may be introduced by your foods, including brine shrimp and especially, tubifex worms. I keep all my dry food in the freezer, except for a small portion that I keep out for feeding purposes. The brine shrimp hatchers and the baster are kept clean with an occasional wash of liquid bleach and a thorough rinsing with hot water. Nets are also disinfected and rinsed after use. If you do have a sick tank, do not transfer any water or material to another tank. If you feel it is necessary to transfer the fish to another tank, do not use one that is freshly set up; use one that has had fish in it before.

Often aquarists will transfer sick fish to a newly set-up tank that has not been conditioned. This immediately compounds their problems; now the fish not only has the

Formalin is one of the oldest fish medications and is still very useful. Photo courtesy of Aquarium Products.

GUPPY HEALTH

There are any number of proprietary medications that can help you cure your fishes. Photo courtesy of Aquarium Products.

disease to contend with but also fresh water, which usually throws them into shock. This cuts their chances of survival by about 50%. I don't wish to imply that all fish are puny, sickly creatures—in fact, some of them you would almost have to hit with a hammer to cause them any discomfort, but you do have to use some common sense to avoid unnecessary problems.

One disease that usually crops up in the female guppy is called columnaris, and looks somewhat like a fungus. It usually manifests itself in spots or lack of pigmentation in the peduncle area of the females, normally on the upper part. If left unchecked, it can spread throughout the tank with a very high kill ratio. It seems bacterial in nature, and from my own observations, is caused by unclean tank or filter conditions. Further manifestation of the disease may be a gradual paralyzing of the body itself. It is also possible that some foods may cause or promote the development of this particular disease. If caught in time this disease can be reversed and cured.

The most effective treatment I have found is

Plistophora is a protozoan infection. It attacks the muscle tissue of the fish.

GUPPY HEALTH 197

Abdominal dropsy. There is no known cure for this condition.

to prepare a dip of 12 drops of 37% formaldehyde (formalin) to a gallon of water. Leave the fish in this solution for 15 to 20 minutes. (Remove them sooner if they show any sign of discomfort.) Remove them from the solution and place them in another conditioned tank which has had two drops of formaldehyde per gallon of water added. Do not feed the first day and in about three days the condition should be cleared up. It is a nasty disease, but one of the easiest to rectify.

The sick tanks should then be taken down and

sterilized with liquid bleach and rinsed well before reusing.

The aforementioned disease should not be confused with ichthyophonus, which is entirely different, even though certain manifestations, such as pigmentation loss, occurs in both. Usually this disease attacks the motive portion of the brain, seemingly affecting males more than females. It becomes noticeable with the males hanging in a vertical rather than a normal horizontal

Set-up for fin surgery.

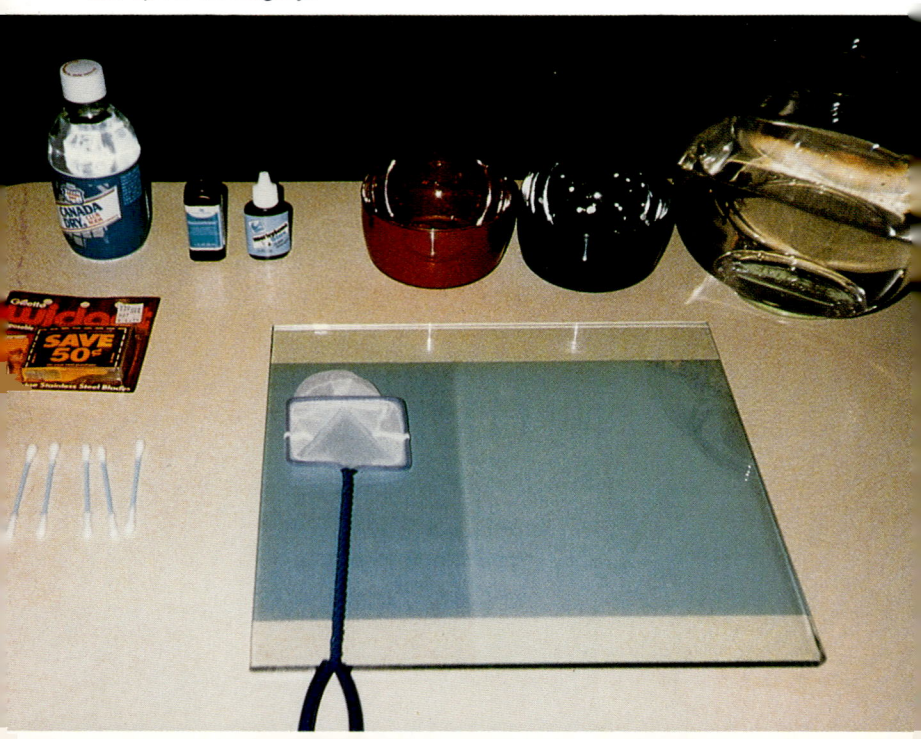

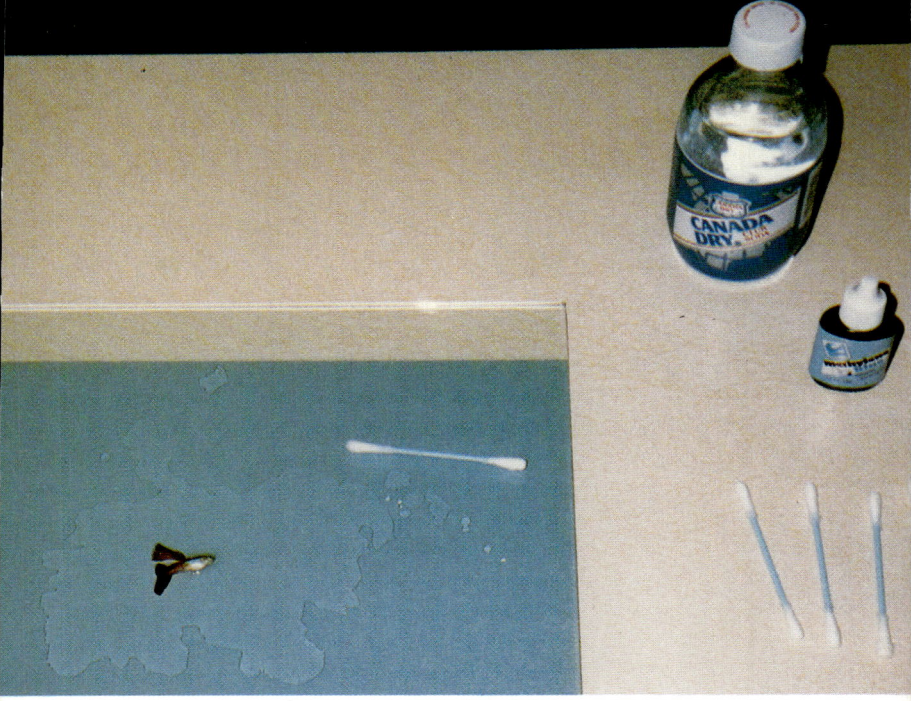

Sometimes if tail rot is not responding to treatment, you can remove the margin of the tail, which will regenerate with healthy new tissue.

position. Upon lightly tapping the side of the tank, fish with disease will swim in a jerking, twirling manner. In a rather short period of time they waste away and die. In a strange way it seems almost hereditary in nature, in that you may have fish from another line in the same tank who do not come down with this disease, while all others from the affected line will die off like flies. Occasionally all the fish in the tank will die.

Occasionally, a fish may develop dropsy, which will cause the body to swell and the scales to protrude. This is not a particularly infectious disease unless the other fish eat the dead fish. Epsom salts can be used to relieve this condition, but unless it is a special fish, it would be simpler to just destroy the fish. This may sound harsh to some, but in many cases it is not worth the effort or time in trying to effect a cure. As is often the case in seriously diseased fish, even though they may pull through they never fully recover and are usually useless for breeding purposes.

Gill rot caused by Branchiomyces.

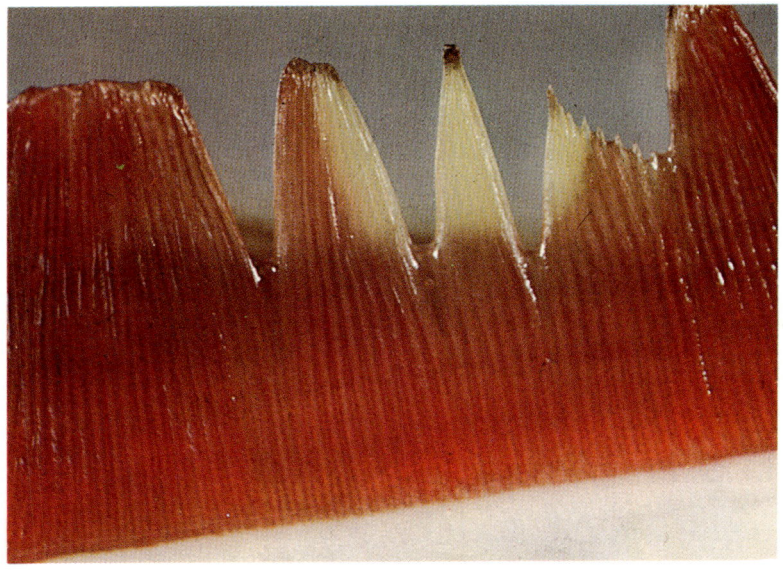

Columnaris *infection with necrosis of the gills.*

Another condition that resembles dropsy is a swelling in the chest area, particularly in males. This is usually caused by fatty deposits within the body cavity itself. In some cases it will develop to the point where the chest actually splits open, causing death. A very rich diet seems to be the culprit. Once the fish reach a certain point it seems irreversible. They continue to remain active and eating until they die. It doesn't seem to be in any way contagious.

With certain diseases it becomes almost necessary to remove the fish from the tank to treat them. For

GUPPY HEALTH

practical purposes, it is not feasible to treat a whole tank of fish when only a couple have problems.

In treating a case of red line tail burn (usually bacterial in nature) the fish should be netted and the affected area treated using a cotton swab dipped in a strong solution of methylene blue or a weak solution of silver nitrate or copper sulfate. Care should be taken that nothing gets into the gill or eye area. In some instances it may be necessary to trim off the infected area of caudal or dorsal by using a razor blade.

Below: *Healthy gills are recognized by their bright red color, free of adhesions.* **Facing Page:** *Congenital deformities of fry. Dispose of any fish that produce this kind of offspring!*

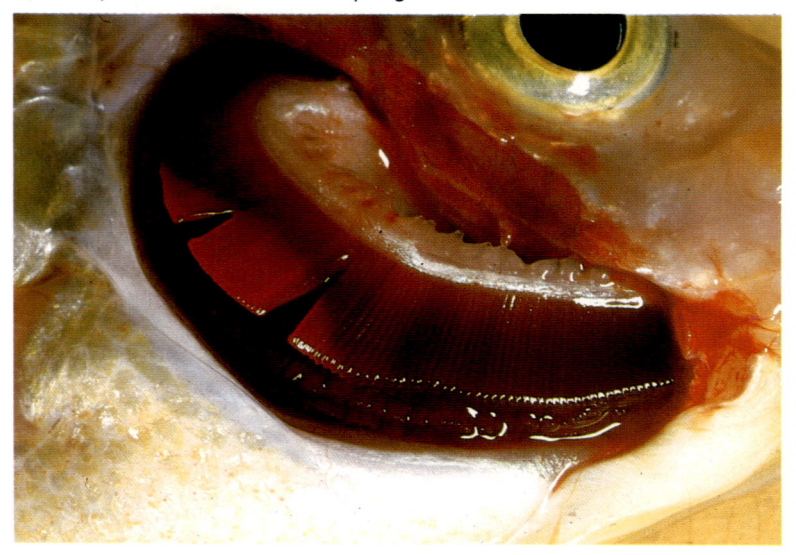

GUPPY HEALTH 203

Occasionally, with certain strains of fish there are genetic weaknesses involved. No matter how ideal your tank conditions are, these fish will develop disease symptoms. If, after a generation or two, no improvement is noted, I would either bring in some new blood for a cross or dispose of these fish.

You can also use ozone generators or ultraviolet light to sterilize the water. However, it is not particularly useful for the guppy exhibitor to use this equipment. Fish raised under these sterile conditions do not do well in exhibitions where they will sometimes be placed in less than an ideal environment. They usually

Cull any fish that show a deformed gill plate.

Gill fluke.

fall apart very quickly. If you were to sell fish raised under these conditions, once placed in a non-sterilized tank set-up, they have almost no immunities and usually expire to whichever disease hits them first.

Stress is a contributing factor in many fish disease problems. Bacteria and some parasites are usually present in all aquarium set-ups, but unless the fish are subjected to stress in some manner, they remain healthy and unaffected.

Occasionally when mixing strains of males in the same tank, One line will get sick while the other fish remain healthy. This would lead one to think that there is an inherited factor involved. In reality it is simply that one line cannot handle a stress situation and falls prey to one of the disease or parasitic conditions present in the tank.

Stress can be caused by a number of factors: the bacteria level could become unbalanced by overfeeding, water quality could become unhealthy due to failure to change water in the proper amount, or the filter could become ineffective due to either too frequent or too infrequent maintenance.

Overfeeding which will cause a bacteria build up

Systemic bacterial infection that is beyond help.

Tapeworms.

can also cause an increase in the ammonia content. Normally if you do not overcrowd your fish and maintain scheduled water changes, ammonia is no problem.

In raising your fish there is a tendency to crowd them somewhat, as you normally start them in a small tank. Also, you may keep them at a slightly higher temperature to increase activity and make them eat more. An additional amount of air to increase the oxygen level in the tank may be used. As long as you maintain the proper balance, things go well. But if, for instance, you should skip a water change or overfeed, you could be in for trouble. First of all you will have additional waste material from the fish which were not removed. Secondly, any uneaten food will increase the ammonia level. The higher temperature will also decrease the oxygen content, which will enhance the ammonia build-up.

With the increased ammonia concentration, the fish's tissue become more sensitive to the intrusion of the bacteria, which in turn can cause fin rot of other secondary infection as well as internal problems. Once this has gotten out of hand it is somewhat hard to rectify, as you wish to continue the growth process. The best solution

Bacterial kidney disease which could easily be mistaken for dropsy.

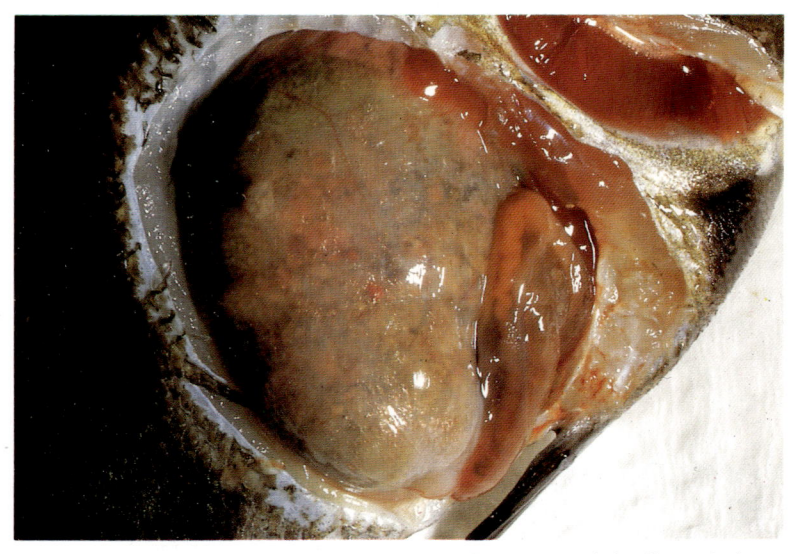

Intestinal obstruction due to inflammation. The stomach is distended with undigested food. Death follows the rupture of the stomach.

is to change the fish to another tank. With older fish, just stop feeding for several days until the tank becomes stabilized.

If there still seems to be a problem, a good treatment to use is four drops of formalin per gallon of water along with one tablespoon of salt per five gallons of water. This combination can be repeated in three days for severe cases. The formalin will dissipate, the salt will not, so a partial water change may be required. Most of the bacteria, both good and bad, will be destroyed with this treatment.

Internal infections, which can cause fish to go off their feed or behave unnaturally due to bacterial infections, can sometimes be helped by antibacterial proprietary medicines.

Some of the dyes such as methylene blue or malachite green are effective if applied directly to the diseased part by a swab. If you were to attempt to treat the entire tank with the required amounts of these medications, everything would be permanently stained. Acriflavine at a 5% solution using two drops per gallon is useful

Bacterial infection of skin wound.

Fungus infection of the mouth, also known as cotton-mouth disease.

for clamped fins or in healing damaged areas. In some instances I have had fair results using vitamin B12 for split fins; sometimes they will heal themselves without using medication, but in most cases they will not. Use 1 cc of water-soluble B12 per five gallons of water.

If a fish displays any abnormal behavior in movement of lack of normal activity, the other fish in the tank will pick on him or her. It is best to remove this fish and place it in a hospital tank or bowl for treatment until it returns to normal activity. Occasionally, an individual fish will become upset for some reason and any movement in front of the tank will cause him to dash to the furthermost portion of the tank away from you. Here he will hover with fins spread and body quivering. The other fish will ignore him, but after a while his behavior will affect them also. Soon after this you will have a tank of frightened, non-eating fish. Just what triggers this reaction in an individual fish is hard to say. It would seem that some fish are like hyperactive humans and any slight stress causes them to flip out, so to speak. It is advisable to

A rupturing cyst of Ichthyoph thirius, *the parasite that causes ich, discharging swarm cells.*

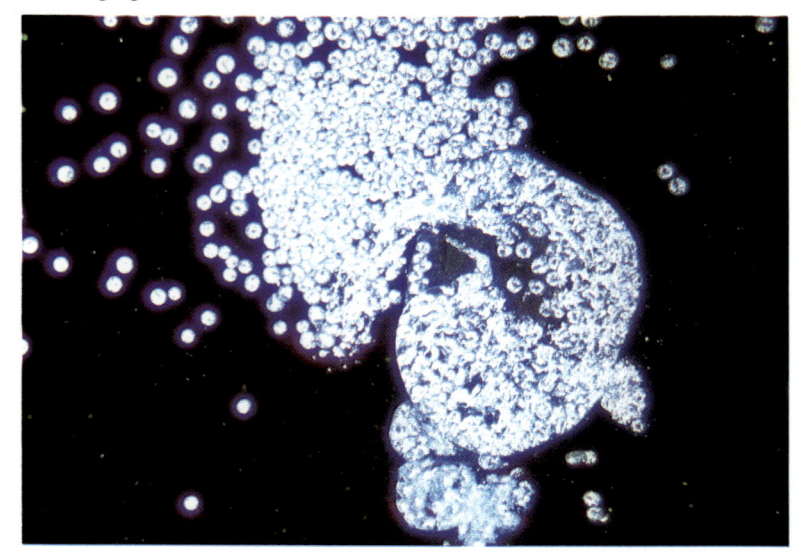

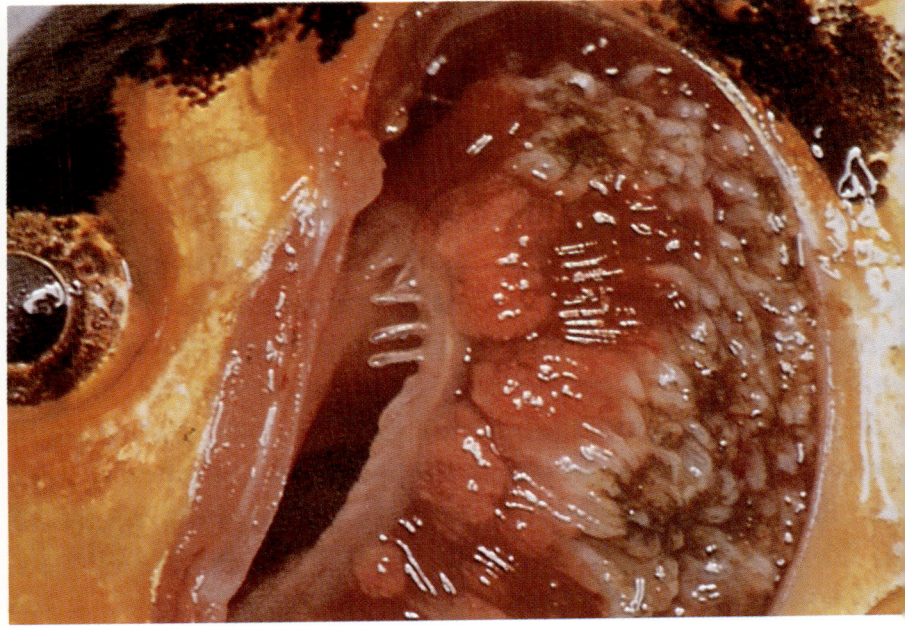

Gill rot, Myxobolus.

move slowly and not make any fast, jerky movements around your fish. When I run into this situation I usually isolate this fish or eliminate him, as he can really mess things up. If more than on fish shows these tendencies, just stop feeding this tank until they wake up and return to normal.

The use of medication under these circumstances usually compounds the problem because of the added stress induced by the addition of a foreign compound to the tank water. Some medications are more harsh than others and cause adverse reactions in the fish under a stress condition.

There are a few other minor diseases that guppies occasionally come down with, but they are rarely epidemic in nature and can usually be treated with over-the-counter medications specific to these diseases.

The fish will usually let you know when they are sick or not feeling well. The main thing to do is keep your eyes open and then take the necessary action to rectify the problem before it gets out of hand.

Tubercular cysts causing distended abdomen.

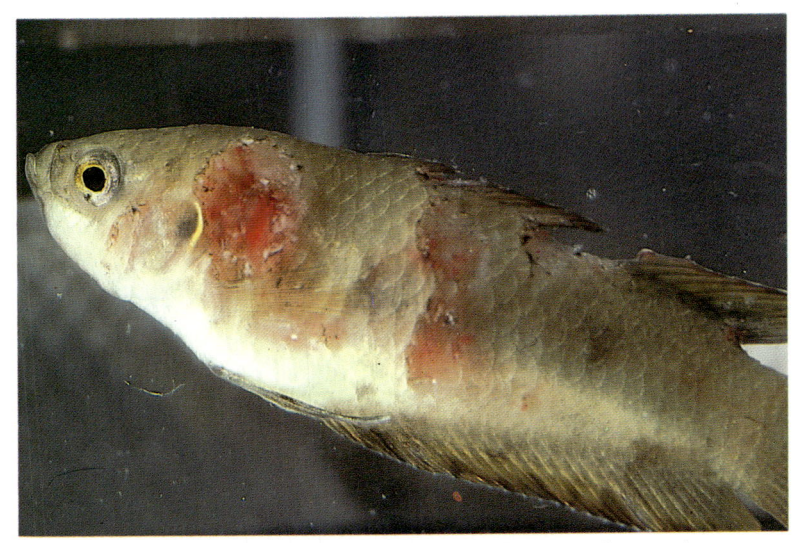

Above: Open sores of tuberculosis. **Below:** Teratoma in a male guppy. A teratoma is a tumor filled with several different kinds of tissue.

A note of caution: all types of medication should be kept up and out of the reach of children. Care should be taken in the use of formalin, even though it is one of the best medications available. Do not get it in your eyes or breathe the fumes directly from the bottle.

I make it a practice to wash well with soap and water after working in the fishroom. Any cuts are treated with merthiolate and so far I haven't had any problems with infections.

All too often in treating our fish we tend to use the headache/aspirin approach. We assume that once a tank has been

Bacterial skin infection following a wound.

Above: *Fancy guppy with* Tetrahymena. *Note the white saddle-like area of depigmentation on the dorsal area of the fish. Fish showing these kinds of lesions usually die within several days.* **Below:** *High-powered photomicrograph of a single* Tetrahymena *protozoan. These small ciliated protozoans are invisible to the naked eye and may number in the hundreds of thousands. The lines in the background are growth rings on the fish scale.*

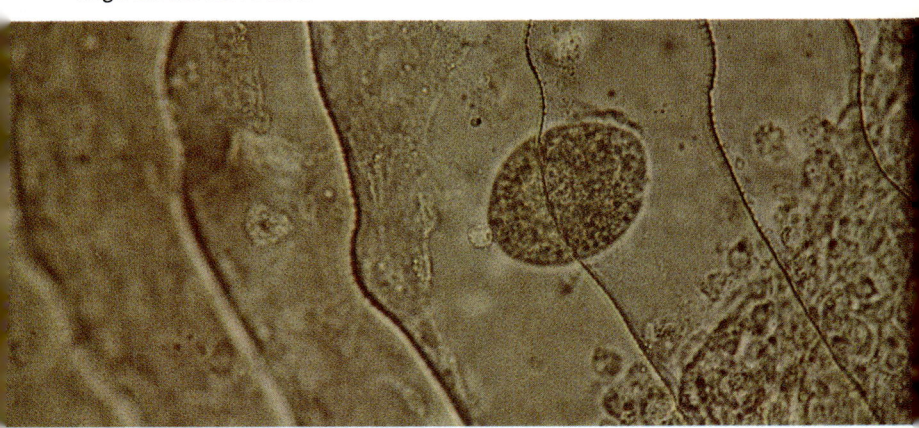

medicated, the fish should be cured in a matter of hours or at least a day or two. It should be realized that some diseases are slow to respond to treatment and therefore it may be necessary to treat the tank several times in order to contain and eliminate the disease.

The diseases with external symptoms respond rather quickly to medications, whereas those that are internal in nature take longer to correct. If it is necessary to isolate a fish for treatment, I would suspend feeding until the condition is cleared up. If it bothers you too much to starve a sick fish, then use only live baby brine shrimp very sparingly. It goes without saying that any dead or dying fish should

Columnaris *infection on skin and scale margins.*

GUPPY HEALTH 219

Above: Severe ich infestation. *Below:* Intestinal flagellate of the genus Bodomonas.

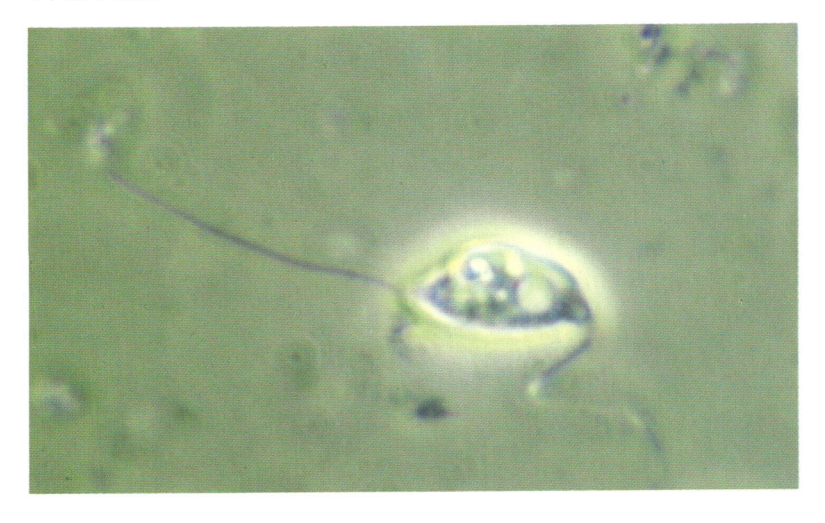

be removed from the tank as soon as noted. This is not for purely aesthetic reasons, but rather because the fish will eat their own dead and dying. While a lot of fish may die from non-transmittable diseases, it is better not to take a risk that the fish will pick up a disease through ingestion.

One disease that has caused quite a bit of trouble in the past few years is the "wasting disease." It usually manifests itself in the following manner: when dry food is added to the tank, the fish swim under the food, make a pass but do not actually take any of the food. If frozen shrimp is placed in the tank, the fish hang suspended in the upper part of the tank pick at the shrimp, but spit it out. Most of the other foods, including live brine

Saprolegnia in guppy.

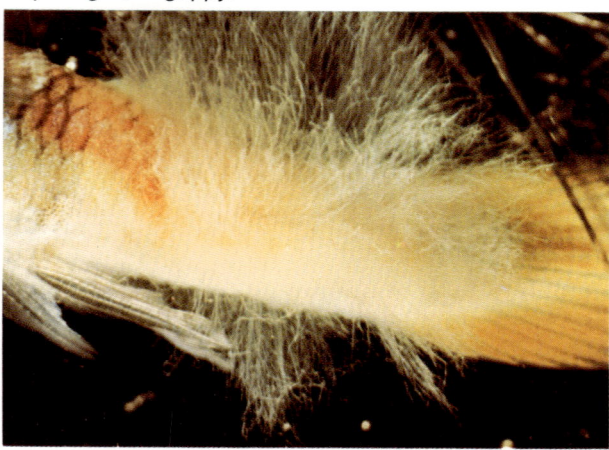

GUPPY HEALTH 221

Tuberculosis with pigment loss and fin necrosis.

shrimp are also rejected. The fish start to slim down and in a matter of a week or so, most of the fish in the tank begin to look like torpedoes. They live for several months in this condition and then they die. Because of the body condition and the length

222 GUPPY HEALTH

of time it takes for the fish to die, it appears that the condition is tubercular in nature.

It has not been a major problem in my tanks, but I have experienced this disease. In most instances, I would try a couple of medications and if they didn't cure the fish completely, I would destroy the fish.

When treating fish for any disease, you have to consider the cost of the

Below: Anchor worms and fish lice (Argulus) are external skin parasites that are easily treated. **Facing Page:** Tuberculosis cysts.

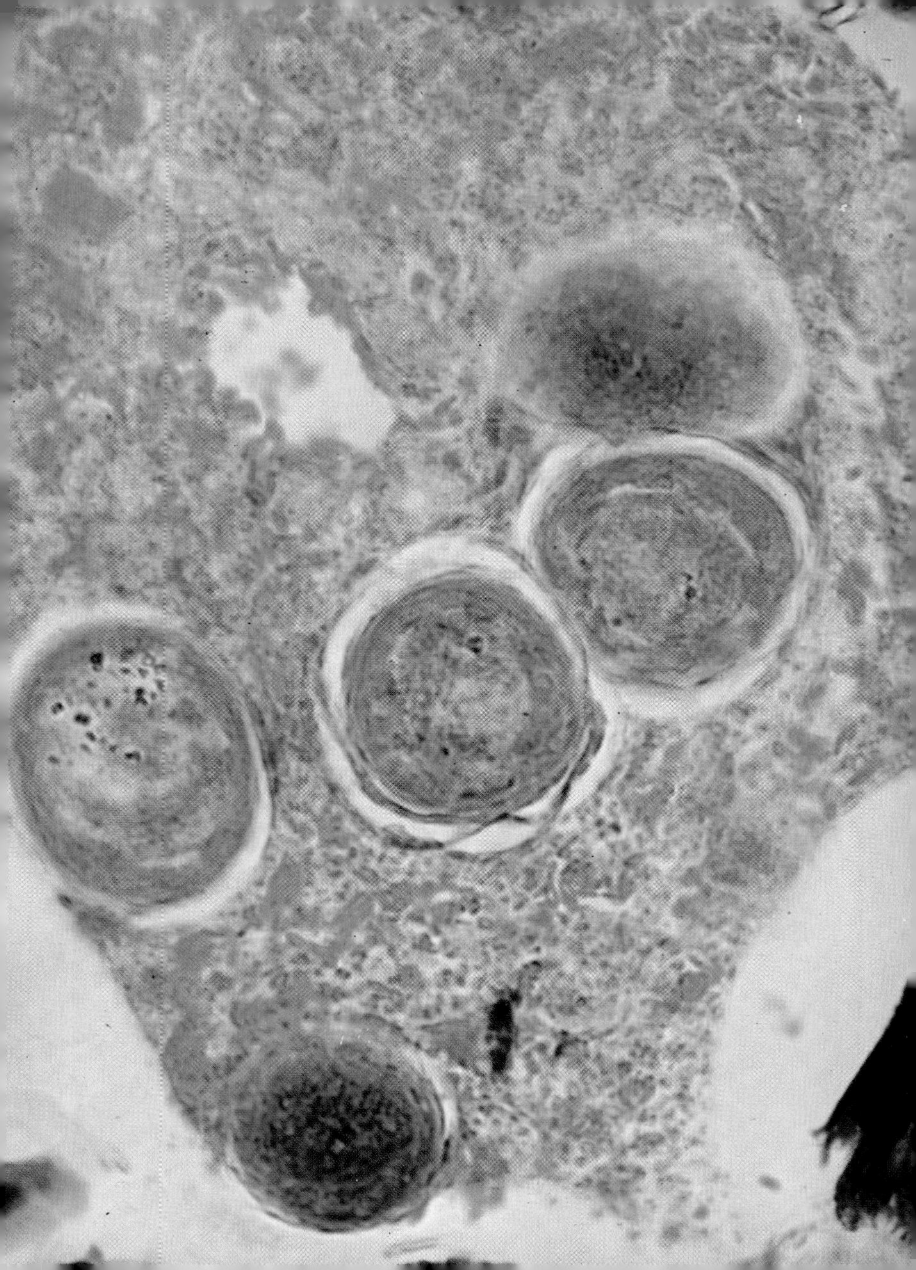

medication, the inconvenience of setting up separate tanks for treatment, etc., and most importantly, whether the fish is going to be any good after it has been cured. Some diseases are easy to treat and the fish is as good as new when

Tail rot with fungus.

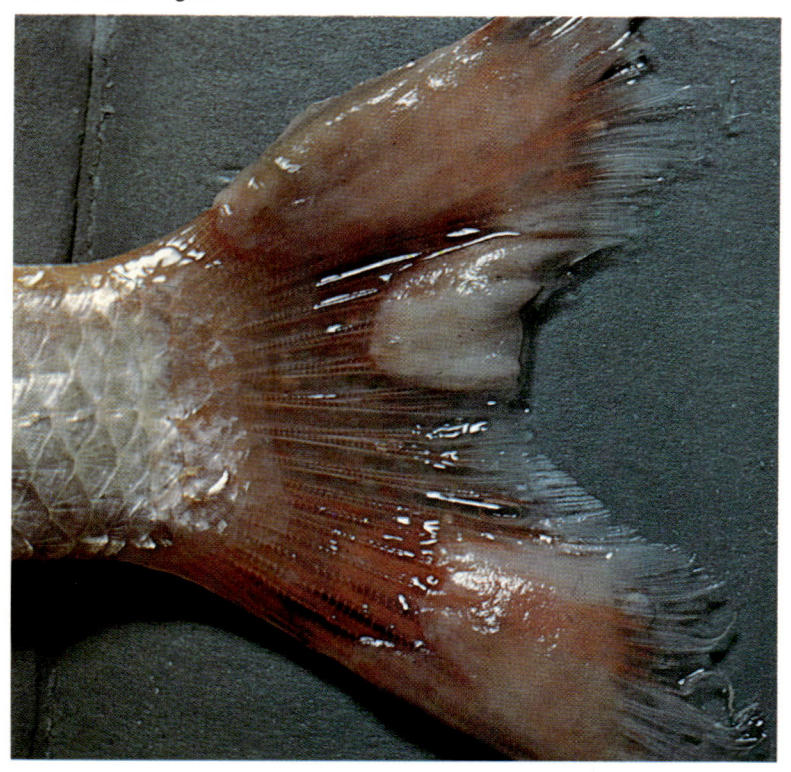

Severe tail rot. So many of these conditions can be prevented with good water quality management!

cured, but other diseases can leave a fish permanently impaired. Most of the time we just guess at our diagnoses. In some instances, it is an educated guess when we have had experience with

Ich.

similar diseases in the past. You can consult fish disease books—one outstanding book is Dieter Untergasser's *Handbook of Fish Diseases*, TFH publication TS-123—for aid in diagnosis and treatments.

Many guppy people have no trouble maintaining healthy tanks because they exercise proper control of their water and filters and practice good feeding habits religiously. I'd like to say that I'm among this number, but it just isn't so. I like to experiment with different ideas too much. Many times it isn't that I don't know any better, but it's sometimes easier to preach than practice.

GUPPY HEALTH 227

Saprolegnia, or fungus. Fungus can kill a fish very quickly so it must be treated without delay.

The small notch in the caudal of this fish is a healed injury and not cause for concern.

Above: *This half black pastel veil has a problem with the trailing edge of his caudal. If left unchecked, it will destroy a good part of his tail. Usually with this type of disease, I will net the afflicted fish out and swab the infected are rather than trying to treat the entire tank, which would disrupt the bacterial balance of the tank.* **Below:** *A nice young half black red. Notice the white spot in the caudal. Every once in a while, tank conditions are not kept up the way they should be and I will end up with some fish with this pigment loss. He will be useless as a show fish, but will be O.K. as a breeder.*

Plants have a mildly antiseptic effect and they remove harmful nitrites and nitrates from the water, thus helping you keep your guppies healthy.

GUPPY HEALTH 231

Variegated snakeskin delta.

Variegated snakeskin delta.

Gold delta.

Above: A few years ago most of the yellow guppies resembled a leopard in the caudal pattern. By selective breeding, yellow deltas such as this one bred by Bob Fillinger are now being shown. Good color match in the dorsal and caudal as well as the body. **Facing Page:** This lovely red veiltail is suffering from a split caudal fin. Perhaps water quality is to blame.

GUPPY HEALTH 233

Multi delta.

Even though showing primary male characteristics, this fish exhibits typical coloration of fancy female guppies.

Guppies are very malleable from a genetic standpoint. The colors and shapes of the tail are fantastic. This is a type of guppy that is produced in Japan.

GUPPY HEALTH 237

This type of guppy is called "grass" in the Orient. They are very popular and can produce some nice specimens. In the U.S., it is called snakeskin.

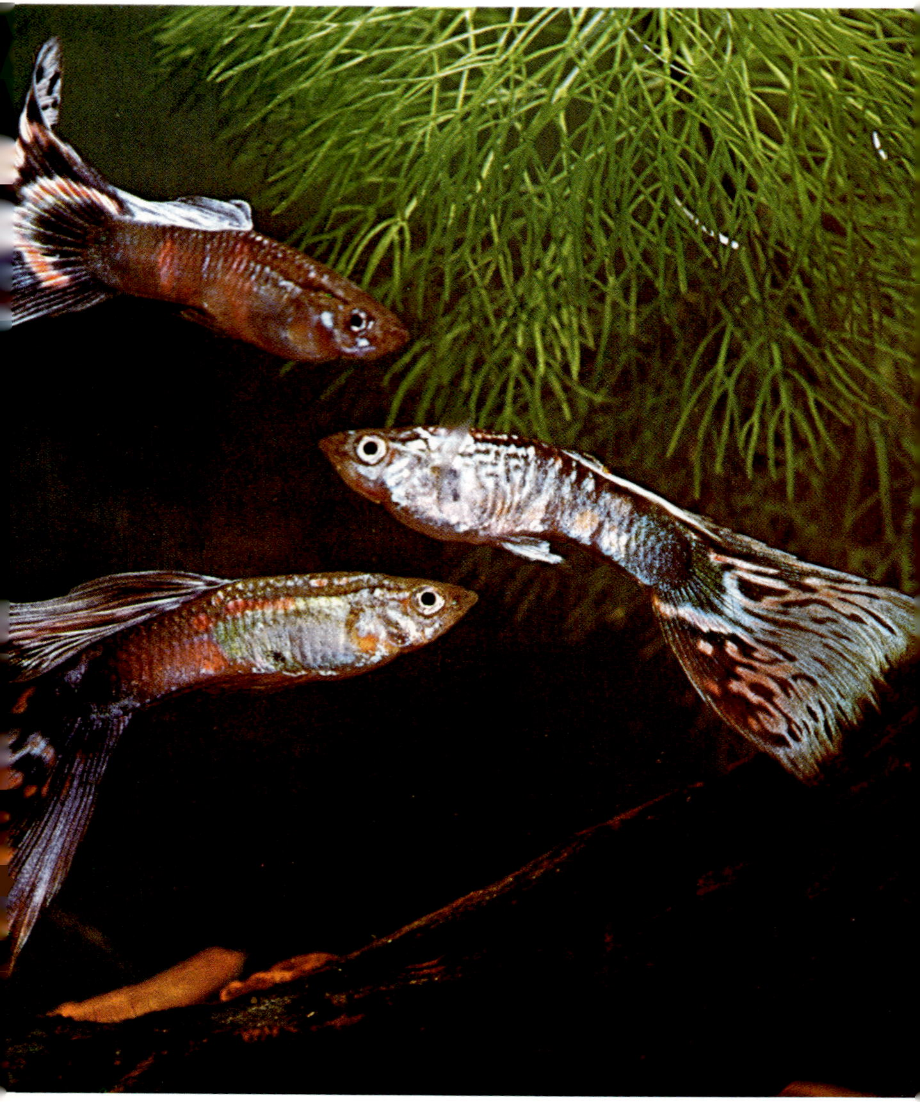

Profit in Raising Guppies

Quite often I am asked, "Is there money to be made in selling guppies?" In all honesty, the answer depends on the type of setup you have and the available market for your fish.

With a small setup of a dozen or so tanks you simply do not have room to keep a lot of fish around for six months or until they are show size. Usually you are forced to dispose of those fish you are not planning to show or to use for breeding. Some of the local pet shops will be happy to take the surplus fish off your hands providing you don't ask too high a price. It's somewhat of a strange situation where some shop owners would rather buy imported fish, even with all the disease problems encountered, than to buy fish from a local breeder.

As a point of illustration: Some years back there was a pet shop in Detroit that sold fish at discount prices. The quality and condition the fish was very poor. Another store a couple of miles away handled good quality, well-acclimated fish but charged more. At that time neon tetras sold for the price of 10 for a dollar at the discount store and twenty nine cents each at the other store, or four for a dollar. (Don't you wish prices were like that now?) Anyway, I ran into a woman at the better store looking at a tank of neons. She

Facing page: Once you start raising guppies for profit, raising guppies becomes a job!

mentioned that she had purchased ten neons last week from the discount store but needed to get some more as nine of them had died. I commented to her that the one neon she had left sure was an expensive fish as it had cost her a dollar. She got a funny look on her face as I moved away. Shortly after that I saw her leaving the store with another bag of neons. I am not inferring that the neons were raised by a local breeder, but rather that many of the stores will not pay a little more in order to get locally raised healthy fish. In very rare cases, because of different water conditions, they may have trouble with local fish also. Moral: Fish are like everything else, you get what you pay for.

Anyway, let's get back to selling fish. With some dealers it is possible to

Multi delta.

Young red breeder showing a good distribution of color in the caudal, dorsal, and body. At six months of age, this is the type of fish I set up for breeding.

make arrangements to trade your fish for supplies at a reduced cost, if not at wholesale prices. In this manner the fish will help pay for the cost of your hobby.

Every few years someone who gets breeding stock from me goes through the initial stages of the small breeder. He is able to sell off his extra fish and maybe another dealer or two will express interest in purchasing fish from him also. The next thing you know he is expanding his setup with another fifty or seventy five tanks or even more. Don't make this mistake. I advise you to go slow and not jump off the deep end. It's far too easy projecting all the money that will be made with the larger set-up than dealing with reality. In a flurry, this breeder buys more breeding stock, a bigger pump, lots of spare filters and other miscellaneous equipment. For the next four to six months he works like crazy getting the fish up to saleable size. The pet shops welcome him with open arms, and two to three weeks later he goes back with more fish. A couple weeks later he calls me and talks about expanding his setup even further to be able to supply the demand. From this point on I just listen and don't offer any further advice as I know what's going to happen.

AOC bicolor. Because of the almost pastel purple and green colors in the caudal, this fish could be classed as a bicolor, although usually there has to be a distinct difference between color or shades of the same color to qualify as a bicolor.

A bronze guppy must have a dark gold body with each scale edged in black such as this one bred by John Magnifico. This delta has a nice pattern in the caudal with a fair dorsal match. Body pattern is well defined.

Sure enough, a few weeks later I receive another call, this time it's to complain about the no-good dealers. Now they want him to cut the prices they were initially paying for the fish in half. Now he has decided not to sell any more fish to these dealers but will go further away to make his sales. He calls back asking if I have any other color fish available, the new dealer wants different types of guppies. He now makes the trip back to get more breeding stock and I notice that he is not quite as enthusiastic about the fish as he once was. He also mentions that he is having some disease problems. Upon questioning him, you learn that he is not cleaning the tanks and changing the water the way he once was, but that's okay because he has figured out a way to take care of the fish without working so hard, and also has found some medication that will clear up his troubles. A few more calls and it is all over. The last call comes asking if I'm interested, or do I know of anyone who might be interested, in buying his tanks and the rest of his equipment.

In a way it's very sad because the guppy hobby has lost a good fish person,

A young male resulting from a hybrid cross between a half black and a solid blue line. He shows a little cobra marking in the front part of the body, along with just a hint of the half black body color.

The colors in the caudal of this multi delta bred by Frank Mormino look almost painted on. A good blend of colors with a fair dorsal match.

but moreso because the individual no longer has the pleasure and relaxation of watching his fish. What people fail to realize is that you can oversaturate a market even in a large metropolitan area in a short period of time. And let's face it, if you are raising fish in quantity you generally try to sell them as soon as possible. No longer do you hold off breeding until you can determine which is the best fish. Within a few generations the fish start to deteriorate in quality. Fish farms can run a profitable operation as they have a nationwide market, and even if the fish are not top quality they can still sell them. With a normal breeder's setup trying to compete with a farm with

Multi delta bred by Frank Mormino. Good looking fish with fair color distribution in caudal and dorsal. Body color is good, showing the three colors to match the caudal. The one-to-one body to caudal ratio is also good. **Facing Page:** *Any young that are born in this tank will be very disappointing. In a very few generations they will have lost any of the features of their parents and be more like the plain wild guppy.*

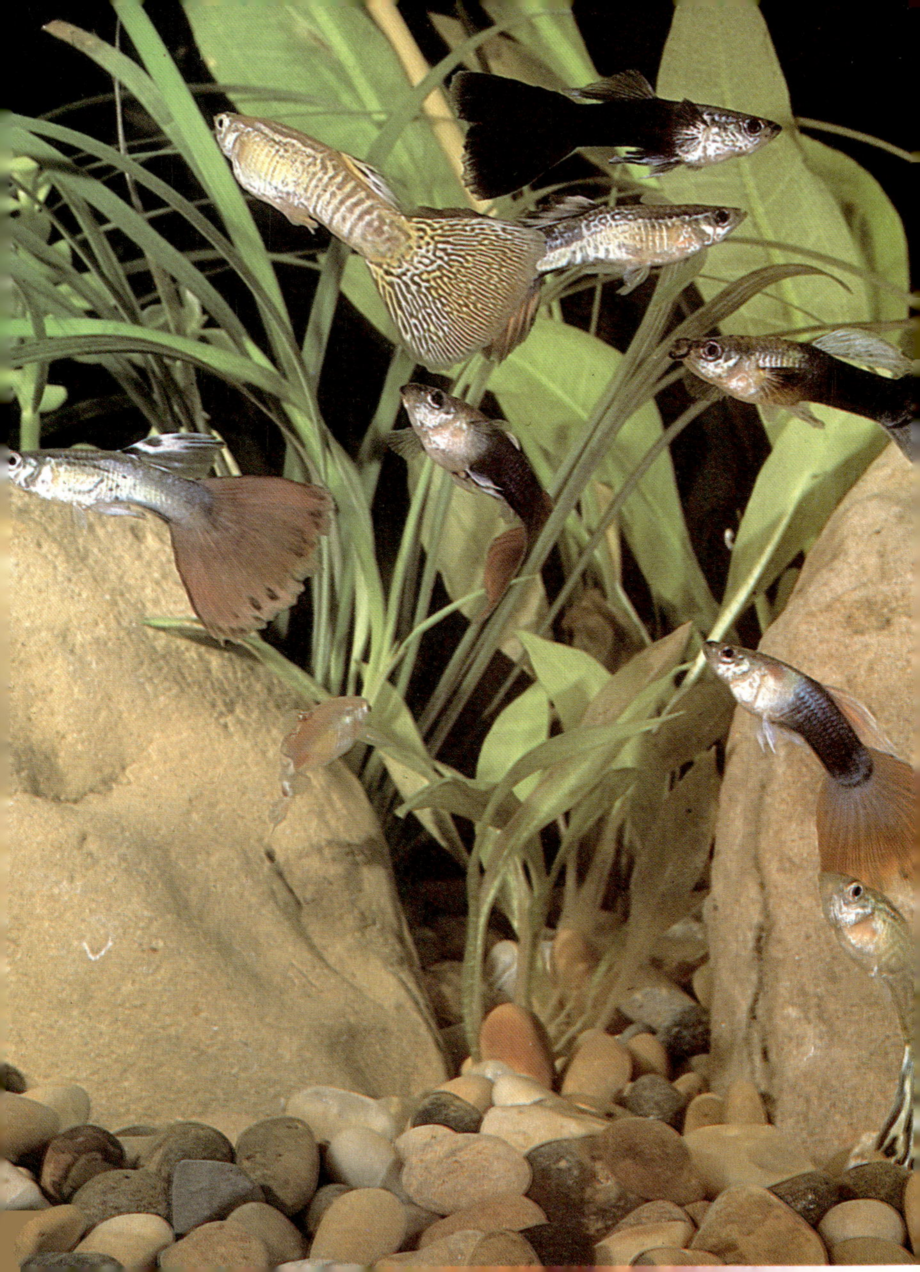

large vats and ponds, to say nothing about the Orient with their cheap labor and food supply, you are fighting an uphill battle.

I have seen individuals who are content to raise what I would call B-grade fish to sell to the shops. This is fine as not everyone is interested in raising show quality fish. These people just like to have fish around. The main thing is to raise your fish and enjoy them.

Many of the top guppy breeders don't even like to sell fish to individuals at all. They would rather dispose of them through wholesalers or a shop. You might well ask, why not advertise and sell your fish direct through the mail rather then sell them for a fraction of the price to a pet shop. The main answer would be complaints. The foremost being that the young fish don't look as good as the parents. People fail to realize that their

Wild lower swordtail.

PROFIT IN RAISING GUPPIES 249

Red lower swordtail.

Pintail.

conditions will not be the same as the breeder's whose fish they have purchased. Consequently, it is unlikely that the fish they raise will reach their full potential, due primarily to the differences in food, water chemistry, and care.

Then in some cases I have sold fish to an individual who, using my stock, raises better fish than I do. This bothered me somewhat at first but then I figured they had more time to devote to the fish than I do, hence the better results.

What it all boils down to is how much effort and time you are willing to put into your fish. Maintaining a couple of dozen tanks and filters along with the associated water changes is not all that much work and can be rather enjoyable. However, when you jump to a hundred or two hundred tanks, there is very little or no relaxation involved; it gets to be pure work.

This female is a relatively new mutation with very elongated pelvic and anal fins.

PROFIT IN RAISING GUPPIES 251

A male of the same mutation with the long pelvic and anal fins.

Another female of the elongated-fin mutation type.

252 SUGGESTED READING

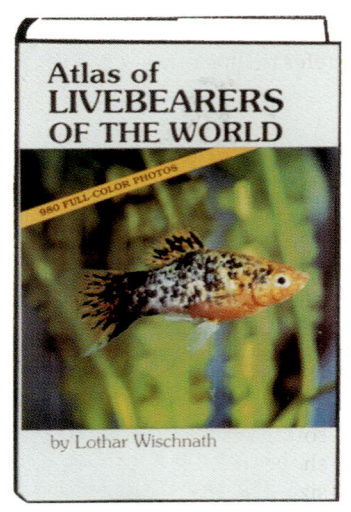

TS-180 336 pages
Over 980 color photos

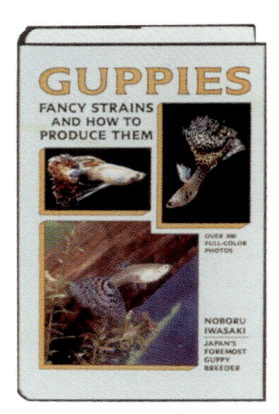

TS-122 144 pages
300 color photos

KW-058 96 pages
60 color photos

SK-035 64 pages
Over 80 color photos

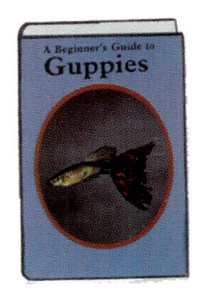

T-117 64 pages
48 color photos

INDEX

Page numbers in **boldface** refer to illustrations.

A
Ab, Dr., 14
Acriflavine, 210
Acrylic aquaria, 26
Air supply, 45
Albino delta guppy, **61**
Albino guppy, **251**
American Guppy Association, 16
Ammonia, 28, 54, 57
Anchor worms, **222**
AOC bicolor delta guppy, **40**, **60**
AOC bicolor guppy, **242**
AOC delta guppy, **129**
AOC guppy, **178**
Argulus, **222**

B
Background, 31
Bacterial
—infection, **206**, 210
—kidney infection, 208
—skin infection, **216**
Baking soda, 79
Beef heart, 91
Bicolor guppy, **235**
Black guppy, 15, **34**
Blackworms, **85**
Bleach, 69
Bloodworm, **85**, **89**
Bloodworms, 79

Blue cobra guppy, **35**
Blue delta guppy, **12**, **19**, **37**, **48**, **52**, **59**, **73**, **131**, **172**
Blue guppy, **13**, **50**, **105**, **109**, **155**, **175**
Blue/green bicolor delta guppy, **39**
Bodomonas, **219**
Body
—color, 179, 183
—condition, 180, 184
—shape, 180, 183
—size, 178, 183
Bottom sword guppy, **10**
Bowl, 27
Branchiomyces, **200**
Breeding, 97, 100
—stock, 103
Breeding tanks, 24
Brine shrimp, **76**, 77, **80**, 81, 83, 194
—baby, 82, 141
—frozen, 83, 92
—hatchers, 77, **78**
—hatching solution, 79
Bronze guppy, **243**

C
Cannibalism, 140
Carbon, **32**, 34, 37, 38, 62
Carotene, 168

INDEX

Caudal
—color, 182, 185
—condition, 182, 185
—shape, 182, 185
—size, 181, 185
Ceratopteris thalicroides, 49
Chlorine remover, 69
Color test, 106
Columnaris, 195, **201**
Columnaris, 218
Congenital deformities, **202**
Congress of Guppy Groups, 19
Cotton-mouth disease, **211**
Culling, 108

D
Daphnia, **86**, 89
Dark blue delta guppy, **21**, **53**
Dark green delta guppy, **39**
Dark purple delta guppy, **171**
Dark purple guppy, **68**
Defects, 127
Deportment, 182, 185
Dominant gene, 112
Dorsal
—color, 180, 184
—condition, 181, 184
—shape, 180, 184
—size, 180, 184
Double sword guppy, **63**
Drip system, 40
Dropsy, **197**

E
Epsom salt, 79

F
Filter
—diatom, 41
—inside box, 34, 37
—media, 38
—outside box, 34
—power, **27**
—sponge, 34, 36
—undergravel, 34, 35
Filtration, 33
—central, 37
Fish lice, **222**
Flake food, **91**
Flamingo guppy, **149**
Formalin, 166, 192, **194**, 197, 209, 216
Fungus, **193**, **224**, **227**

G
Gill fluke, **205**
Gill rot, **200**, **213**
Glass aquaria, 25
Gold albino delta guppy, **43**
Gold delta guppy, **231**
Gold red-tailed delta guppy, **149**
Gold snakeskin delta guppy, **174**
Gonopodium, 116, **117**, 142
Gordon's Formula, 90
Grass guppy, **237**
Gravid spot, **120**, 142
Gray-bodied half black pastel AOC guppy, **165**
Gray-bodied half black guppy, **104**
Gray-bodied half black red guppy, **107**
Green delta guppy, **5**, **18**, **45**, **176**, **177**
Guppy, 13
—wild lower swordtail, **248**
—wild-type, **11**, **118**

H

Hahnel, Paul, **9**, 10, 14
Half black AOC delta, **161**
Half black AOC delta guppy, **150**
Half black AOC guppy, **101**, **157**, **164**
Half black blue delta guppy, **130**, **163**
Half black blue guppy, **18**, **23**
Half black delta guppy, **51**
Half black pastel delta guppy, **148**
Half black pastel veiltail guppy, **129**
Half black red delta guppy, **71**, **75**
Half black red gold guppy, **170**
Half black red guppy, **18**, **21**, **97**, **100**, **122**, **127**, **229**
Half black red veil guppy, **64**
Half black yellow delta guppy, **167**
Hardness, 58
Health, 187
Heater, **36**
Heating, 41
Hybrid vigor, 117

I

Ich, **192**, 212, **219**, **226**
Ichthyophorus, 198
Ichthyophythirius, **212**
I.F.G.A. Point System, 179
Inbreeding, 108
International Fancy Guppy Assocation, 16, 19, 159
Intestinal obstruction, **209**

J

Judging, 160, 177
Judging blue book, 178

K

Kaufman, Henry, 10
Konig, Larry, 10

L

Lethal gene, 123
Light purple guppy, **66**
Lighting, 31
—fluorescent, 31
—incandescent, 31
—nite-light, 31
Linebreeding, 108, 114
Longevity, 149
Longfin multi guppy, **152**
Lordosis, **188**

M

Malachite green, 210
Marbles, 38
Medium blue delta guppy, **69**
Methylene blue, 210
Mosquito, **89**
Mosquito larva,
Mule, 155
Multi delta guppy, **145**, **153**, **154**, **234**, **240**, **245**, **246**
Multi veil guppy, **47**
Multi-colored delta guppy, **133**
Myxobolus, **213**

N

Nitrate, 57

O

Outcrossing, 108, 117
Overfeeding, 89
Oxygen, 73

P

Paprika, 168

pH, 54, 56, 58
Pingu®, **74**
Pintail guppy, **139**, **141**, **249**
Plants, **230**
Plistophora, **196**
Plistophora hyphessobryconis, **187**
Poecilia reticulata, **11**, 13
Prepared foods, 84
Pump, 66, 67
Purple delta guppy, 33, **41**

R
Rack, 28
Recessive gene, 112
Red albino delta guppy, **44**
Red bicolor delta guppy, **135**
Red delta guppy, **42**, **62**, **72**, **160**, **168**
Red delta R x R line, **57**
Red guppy, 16, **48**, **241**
Red line tail burn, 202
Red lower swordtail, **249**
Red snakeskin guppy, **143**
Red veil guppy, **58**, **70**
Red veiltail guppy, **233**

S
Salt, 69, 209
Saprolegnia, **193**, **220**, **227**
Shipping, 163
Shipping box, **162**
Show classes, 168
Show tanks, 22
Showing, 159
Shubel, Stan, **6**

Snakeskin guppy, **8**, **35**, **151**, **237**
Solar salt, 79
Solid snakeskin guppy, **67**
Sternke, William, 10, 14
Symmetry, 182, 185

T
Tail rot, **190**, 191, **199**, **224**, **225**
Tapeworm, **207**
Teratoma, **215**
Testosterone, 105
Tetracycline, 166
Tetrahymena, **217**
Thermometer, **36**
Top sword guppy, 15
Top sword pintail guppy, **133**
Tubercular cysts, **214**
Tuberculosis, **215**, **221**, 222
Tubifex worms, **87**, 88, 92

V
Variegated snakeskin delta guppy, **231**
Variegated snakeskin guppy, **47**, **65**
Vitamin B12, 211

W
Water changes, 57, 64
Water conditioners, 54
Water quality, 52
Water sprite, 49
Whiteworms, 89

Y
Yellow delta guppy, **232**